ILLUSTRATED DICTIONARY

for

ELECTRICAL WORKERS

John Traister

Delmar Publishers' Online Services
To access Delmar on the World Wide Web, point your browser to:
http://www.delmar.com/delmar.html
To access through Gopher: gopher://gopher.delmar.com
(Delmar Online is part of "thomson.com", an Internet site with information on
more than 30 publishers of the International Thomson Publishing organization.)
For information on our products and services:
email: info@delmar.com
or call 800-347-7707

Delmar Publishers Inc.®

NOTICE TO THE READER

Publisher does not warrant or guarantee any of the products described herein or perform any independent analysis in connection with any of the product information contained herein. Publisher does not assume, and expressly disclaims, any obligation to obtain and include information other than that provided to it by the manufacturer.

The reader is expressly warned to consider and adopt all safety precautions that might be indicated by the activities described herein and to avoid all potential hazards. By following the instructions contained herein, the reader willingly assumes all risks in connection with such instructions.

The publisher makes no representations or warranties of any kind, including but not limited to, the warranties of fitness for particular purpose or merchantability, nor are any such representations implied with respect to the material set forth herein, and the publisher takes no responsibility with respect to such material. The publisher shall not be liable for any special, consequential or exemplary damages resulting, in whole or in part, from the readers' use of, or reliance upon, this material.

Cover design by Elana Suzanne

Delmar Staff
New Product Acquisitions: Mark W. Huth
Project Editors: Carol Micheli, Andrea Edwards Myers
Production Coordinator: Wendy Troeger
Art Coordinator: Michael Nelson

For information, address Delmar Publishers Inc.
3 Columbia Circle, PO Box 15015
Albany, New York 12212–5015

Printed in the United States of America
published simultaneously in Canada
by Nelson Canada,
a division of the Thomson Corporation

10 9 8 7 6

Library of Congress Cataloging-in-Publication Data

Traister, John E.
  Illustrated dictionary for electrical workers/John Traister.
    p. cm.
  ISBN 0-8273-4763-4
  1. Electrical engineering-Dictionaries I. Title.
  TK9.T73 1991
  621.3'03--dc20   90-28289
                                                                   CIP

Preface

This *Illustrated Dictionary for Electrical Workers* has been prepared for the purpose of assisting students, apprentice electricians, electrical designers, and other workers in securing an understanding of the technical terms with which they come in daily contact.

While many technical books contain glossaries of the terms mentioned in their texts, the authors, in many cases, assume that those interested should be so familiar with the terms described that definitions should not be needed. They therefore proceed with their descriptions of the application, instead of prefacing their remarks with understandable definitions.

Definitions of many of the terms listed in this book will not be found in the average dictionary or in any technical text even though they are, almost with exception, terms that are used in electrical construction projects throughout the United States and Canada.

A great effort has been made to keep the definitions simple, yet thorough. The many illustrations further assist the reader in understanding more than 1000 terms.

The main purpose of this book is to give the worker a knowledge of trade nomenclature that will be extremely useful in the pursuit of his or her vocation.

John Traister
Bentonville, Virginia
1991

A

AA (Aluminum Association): A manufacturers' association that promotes the use of aluminum.

AAC: All aluminum conductor.

AASC: Aluminum alloy stranded conductors.

abrasion: The process of rubbing, grinding, or wearing away by friction.

abrasion resistance: Ability to resist surface wear.

abrasive paper: Paper or cloth on which flint, garnet, emery, aluminum oxide, or corundum has been fastened with glue or some other adhesive. One use in the electrical field is to clean conductors, contacts, or terminals.

ac (alternating current): 1) A periodic current, the average of which is zero over a period; normally the current reverses after given time intervals and has alternately positive and negative values. 2) The type of electrical current actually produced in a rotating generator (alternator).

accelerated life tests: Subjecting a product to operating conditions more severe than normal to expedite deterioration, affording some measure of probable life at normal conditions.

accelerator: 1) A substance that increases the speed of a chemical reaction. 2) Something to increase velocity.

acceptable (nuclear power): Demonstrated to be adequate by the safety analysis of the station.

acceptance test: Made to demonstrate the degree of compliance with specified requirements.

accepted: Approval for a specific installation or arrangement of equipment or materials.

accessible: Capable of being removed or exposed without damaging the building structure or finish, or not permanently closed in by the structure or finish of the building.

ACSR (aluminum, conductor, steel reinforced): A bare composite of aluminum and steel wires, usually aluminum around steel.

actuated equipment (nuclear power): Component(s) that perform a protective function.

administrative authority: An organization exercising jurisdiction over the National Electrical Safety Code.

AEC (Atomic Energy Commission): Now defunct; see *ERDA* and *NRC*.

AEIC: Association of Edison Illuminating Companies.

aggregate: Material mixed with cement and water to produce concrete.

aging: The irreversible change of material properties after exposure to an environment for an interval of time.

AIA: 1) American Institute of Architects. 2) Aircraft Industries Association.

air cleaner: Device used for removal of airborne impurities.

air diffuser: Air distribution outlet designed to direct airflow into desired patterns. See Fig. A-1.

air entrained concrete: Concrete in which a small amount of air is trapped by addition of a special material to produce greater durability.

air oven: A lab oven used to heat by convection of hot air.

Al: Aluminum.

Al-Cu: An abbreviation for aluminum and copper, commonly marked on terminals, lugs, and other electrical connectors to indicate that the device is suitable for use with either aluminum conductors or copper conductors.

alive: Energized; having voltage applied.

alligator wrench: A wrench with toothed V-shaped jaws fixed in position.

alloy: A substance having metallic properties and being composed of elemental metal and one or more chemical elements.

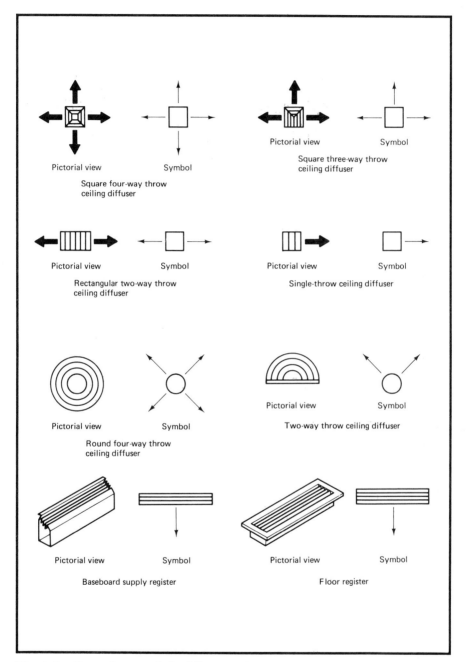

Fig. A-1: Several types of air diffusers. Also shown are two types of supply registers for comparison.

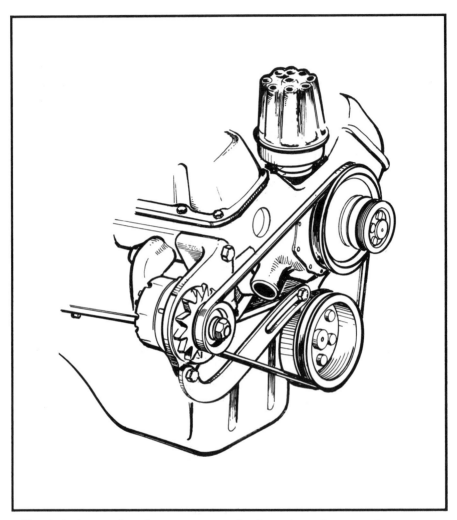

Fig. A-2: Automotive alternators are used to keep the car's storage battery charged.

alternator: A device to produce alternating current. Alternators range in size from small automotive types (Fig. A-2) to huge types used in power plants to produce electricity for cross-country distribution, as shown in Fig. A-3.

Alumoweld®: An aluminum clad steel wire by Copperweld Steel Corp.

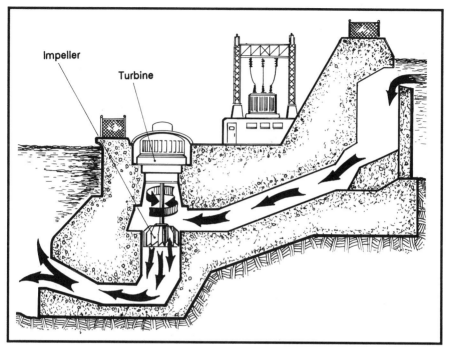

Fig. A-3: Large hydroelectric power plants supply huge amounts of our electric energy.

ambient temperature: Temperature of fluid (usually air) that surrounds an object on all sides.

American bond: Brickwork pattern consisting of five courses of stretchers followed by one bonding course of headers.

ammeter: An electric meter used to measure current, calibrated in amperes.

ampacity: The current-carrying capacity of conductors or equipment, expressed in amperes.

ampere (A): The basic SI unit measuring the quantity of electricity.

ampere-turn: The product of amperes times the number of turns in a coil.

amplification: Procedure of expanding the strength of a signal.

amplifier: 1) A device that enables an input signal to directly control a larger energy flow. 2) The process of increasing the strength of an input.

amplitude: The maximum value of a wave.

analog: Pertaining to data from continuously varying physical quantities.

angle bracket: A form of support having two faces generally at right angles to each other. A web is often added to increase strength.

angle, roll over (overhead): The sum of the vertical angles between the conductor and the horizontal on both sides of the traveler; excessive roll over angles can cause premature splice failures.

angular velocity: The average time rate of change of angular position; in electrical circuits = 2f, and f equals frequency.

ANI (American Nuclear Insurers): A voluntary unincorporated association of companies providing property and liability insurance for US nuclear power plants; formerly *NELPIA.*

annealing: The process of preventing or removing objectional stresses in materials by controlled cooling from a heated state; measured by tensile strength.

annealing, bright: Annealing in a protective environment to prevent discoloration of the surface.

anode: 1) Positive electrode through which current enters a non-metallic conductor such as an electrolytic cell. 2) The negative pole of a storage battery.

ANSI (American National Standards Institute): An organization that publishes nationally recognized standards.

antenna: A device for transmission or reception of electromagnetic waves.

antioxidant: Retards or prevents degradation of materials exposed to oxygen (air) or peroxides.

antisiphon trap: Trap in a drainage system designed to preserve a water seal by defeating siphonage.

aperture seal (nuclear): A seal between containment aperture and the electrical penetration assembly.

appliance: Equipment designed for a particular purpose, using electricity to produce heat, light, mechanical motion, etc.; usually complete in itself, generally other than industrial use, normally in standard sizes or types.

approved: 1) Acceptable to the authority having legal enforcement. 2) Per *OSHA*: A product that has been tested to standards and found suitable for general application, subject to limitations outlined in the nationally recognized testing lab's listing.

apron: Piece of horizontal wood trim under the sill of the interior casing of a window.

arc: A flow of current across an insulating medium.

arc furnace: Heats by heavy current flow through the material to be heated.

arcing time: The time elapsing from the severance of the circuit to the final interruption of current flow.

arc resistance: The time required for an arc to establish a conductive path in or across a material.

areaway: Open space below the ground level immediately outside a building. It is enclosed by substantial walls.

armature: 1) Rotating machine: the member in which alternating voltage is generated. 2) Electromagnet: the member that is moved by magnetic force. See Fig. A-4.

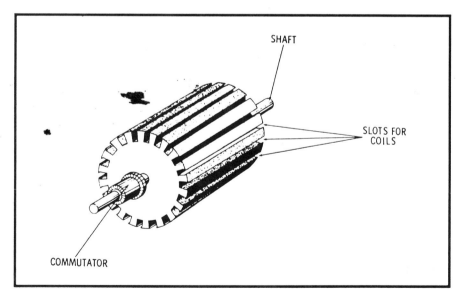

SHAFT

SLOTS FOR
COILS

COMMUTATOR

Fig. A-4: The armature is a moving part in all electric motors and generators.

armor: Mechanical protector for cables; usually a helical winding of metal tape, formed so that each convolution locks mechanically upon the previous one (interlocked armor); may be a formed metal tube or a helical wrap of wires. See Fig. A-5.

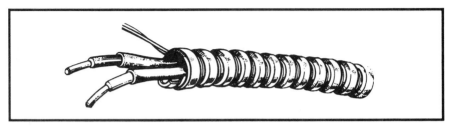

Fig. A-5: Type AC (armor-clad) cable. The interlocking metal covering is designed to protect the cable conductors.

arrester: Wire screen secured to the top of an incinerator to confine sparks and other products of burning.

ashlar: Squared and dressed stones used for facing a masonry wall; short upright wood pieces extending from the attic floor to the rafters forming a dwarf wall.

askarel: A synthetic insulating oil that is nonflammable but very toxic—being replaced by silicone oils.

ASME: American Society of Mechanical Engineers.

associated circuits (nuclear power): Nonclass 1E circuits that share power supplies or are not physically separated from Class 1E circuits.

ASTM (American Society for Testing and Materials): A group writing standards for testing materials and specifications for materials.

asymmetrical: Not identical on both sides of a central line; unsymmetrical.

atom: According to our present understanding, the atom is believed to consist of a central nucleus composed of protons and neutrons, surrounded by orbiting electrons, as shown in Fig. A-6. The nucleus is relatively large when compared with the orbiting electrons, the same as our sun is large when compared to its orbiting planets; also, the orbiting satellites are small in comparison to the satellites' planet.

The orbiting electrons are held in place by the attractive electric force between the electron and the nucleus—similar to how the earth's gravity keeps its satellite (the moon) from drifting off into space. The law of charges states that opposite charges attract while like charges repel. The positive charge of the protons in the nucleus, therefore, attracts the negatively-charged electrons.

If this force of attraction were the only one in force, the electrons would be pulled closer and closer to the nucleus and eventually be absorbed into the nucleus. However, this force of attraction is balanced by the centrifugal force that results from the motion of the electrons around the nucleus.

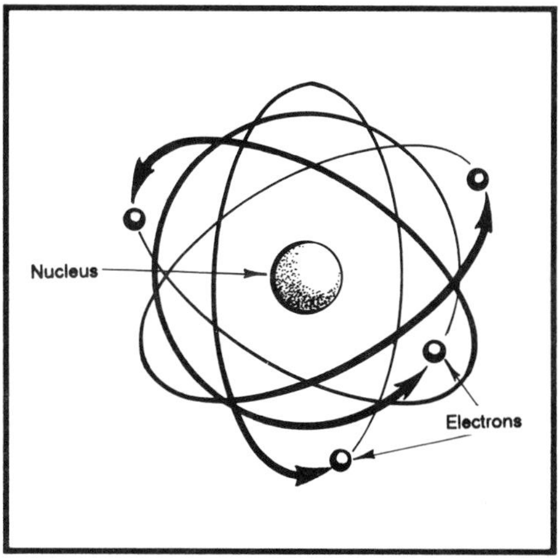

Fig. A-6: An atom is believed to consist of a central nucleus composed of protons and neutrons, surrounded by orbiting electrons.

attachment plug or cap: The male connector for electrical cords. Fig. A-7 gives the most common types of receptacles for use with various types of plugs.

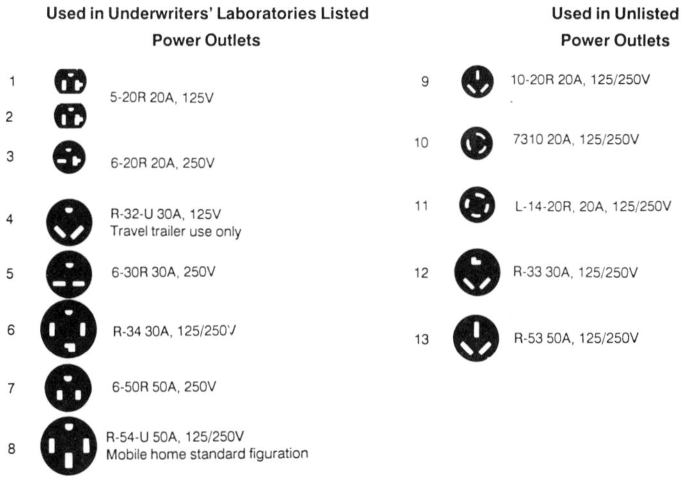

Used in Underwriters' Laboratories Listed Power Outlets		Used in Unlisted Power Outlets	
1	5-20R 20A, 125V	9	10-20R 20A, 125/250V
2			
3	6-20R 20A, 250V	10	7310 20A, 125/250V
4	R-32-U 30A, 125V Travel trailer use only	11	L-14-20R, 20A, 125/250V
5	6-30R 30A, 250V	12	R-33 30A, 125/250V
6	R-34 30A, 125/250V	13	R-53 50A, 125/250V
7	6-50R 50A, 250V		
8	R-54-U 50A, 125/250V Mobile home standard figuration		

Fig. A-7: Receptacle configurations.

attenuation: A decrease in energy magnitude during transmission.

audible: Capable of being heard by humans.

auditable data: Technical information that is documented and organized to be readily understandable and traceable to independently verify inferences or conclusions based on these records.

auger: A wood-boring tool of large size with handle attached at right angles to the tool line. Several types are made for different purposes.

autoclave: A heated pressure vessel used to bond, cure, seal, or used for environmental testing.

automatic: Operating by own mechanism when actuated by some impersonal influence; nonmanual; self-acting.

automatic transfer equipment: A device to transfer a load from one power source to another, usually from normal to emergency source and back.

autotransformer: Any transformer where primary and secondary connections are made to a single cell. The application of an autotransformer is a good choice for some users where a 480Y/277- or 208Y/120-volt, three-phase, four-wire distribution system is utilized.

auxiliary: A device or equipment that aids the main device or equipment.

AWG (American Wire Gage): The standard for measuring wires in America.

awl: A small pointed tool for making holes for nails or screws. When used to mark metal objects, it is sometimes called "scratch awl."

B

backfill: Loose earth placed outside foundation walls for filling and grading; fill of broken stone or other coarse material outside a foundation or basement wall to provide drainage.

back pressure: Pressure in the low side of a refrigerating system; also called suction pressure or low side pressure.

balanced circuit: An electrical circuit so adjusted with respect to nearby circuits as to escape the influence of mutual induction; a three-wire circuit having the same load on each side of the neutral wire.

ballast: A device designed to stabilize current flow.

balloon framing: System of small house framing; two by fours extending two stories with inch by quarter ledger strips notched into the studs to support the second-story floor beams. See Fig. B-1 on page 16.

bank: An installed grouping of a number of units of the same type of electrical equipment; such as "a bank of transformers" or "a bank of capacitors" or a "meter bank," etc.

bar: A long, solid product having one cross-sectional dimension of 0.375 inch or more.

bare (conductor): Not insulated; not coated.

bargeboard: Ornamented board covering the roof boards and projecting over the slope.

bar magnet: 1) A straight permanent magnet. 2) A permanent magnet made in a bar shape.

barometer: Instrument for measuring atmospheric pressure.

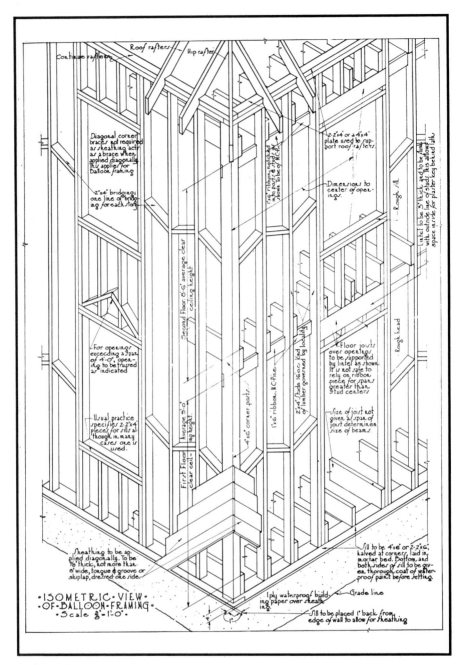

Fig. B-1: Residential building using balloon framing.

barrier: A partition, such as an insulating board to separate bus bars of opposite voltages.

base: One of the regions or terminals of a transistor.

base ambient temperature: The temperature of a cable group when there is no load on any cable of the group or on the duct bank containing the group.

baseboard: The finishing board that covers the plaster wall where it meets the floor.

base line: A definite line from which measurements are taken in the layout of work.

base load: The minimum load over a period of time.

batten: Narrow wood strips used to cover joints.

batter: Slope of the exposed face of a retaining wall.

battery: A device that converts chemical to electrical energy; used to store electricity. See Fig. B-2.

battery acid: The acid used in forming the electrolyte of a storage battery. Sulphuric acid is commonly used.

battery capacity: The number of ampere hours that can be obtained from a storage battery.

battery charger: An electrical device for charging a battery. It usually converts alternating current to direct current for charging purposes.

battery container: The hard rubber or plastic composition casing into which the elements and the electrolyte are placed.

bayonet socket: A lamp socket having two lengthwise slots in its sides. These slots take a right-angled turn at the bottom, so that a lamp with two pins may be connected by pushing it into the socket and giving it a slight turn. See Fig. B-3 on page 18.

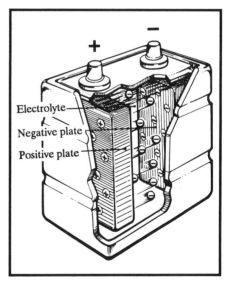

Electrolyte
Negative plate
Positive plate

Fig. B-2: Typical storage battery.

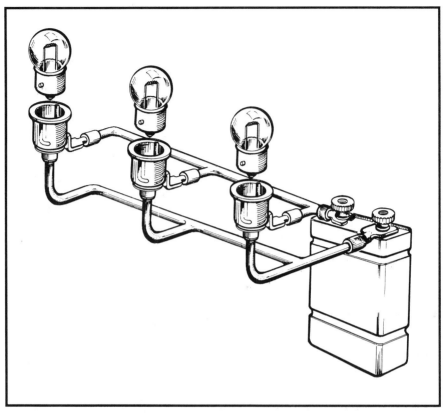

Fig. B-3: Three bayonet lamp sockets connected in parallel.

bead: Narrow projecting molding with a rounded surface; or in plastering, a metal strip embedded in plaster at the projecting corner of a wall.

beam: A horizontal member of wood, reinforced concrete, steel, or other material used to span the space between posts, columns, girders, or over an opening in a wall.

bearing plate: Steel plate placed under one end of a beam or truss for load distribution.

bearing wall: Wall supporting a load other than its own weight.

bed: Place or material in which stone or brick is laid; horizontal surface of positioned stone; lower surface of brick.

bedding: A layer of material to act as a cushion or interconnection between two elements of a device, such as the jute layer between the sheath and wire armor in a submarine cable; sometimes incorrectly used to refer to extruded insulation shields.

belt: The outer protective nonmetallic covering of cable; its jacket.

belted type cable: A multiple conductor cable having a layer of insulation over the core conductor assembly.

bench mark: Point of reference from which measurements are made.

Bessel function: A mathematical solution to a differential equation that is used to solve changes in conductor resistance and mutual inductance between conductors with respect to frequency changes due to skin and proximity effects.

bias vacuum tube: Difference of potential between control grid and cathode. Transistor—difference of potential between base and emitter and base and collector. Magnetic amplifier—level of flux density in the magnetic amplifier core under no-signal condition.

bid: A proposal to furnish supplies or equipment; to carry out or to perform certain work for a specified sum.

BIL (Basic Impulse Level): A reference impulse insulation strength.

bimetal strip: Temperature regulating or indicating device that works on the principle that two dissimilar metals with unequal expansion rates, welded together, will bend as temperature changes.

binder: Material used to hold assembly together.

bipolar: Having two magnetic poles of opposite polarity.

birdcage: The undesired unwinding of a stranded cable.

blackbody: A hypothetical body that absorbs, without reflection, all of the electromagnetic radiation incident on its surface.

blister: A defect in metal, on or near the surface, resulting from the expansion of gas in the subsurface zone. Very small blisters may be called "pinheads" or "pepper blisters."

block bridging: Solid wood members nailed between joists to stiffen a floor.

blueprint: See *drawing, electrical.*

bobbins: Metal spools; small metal reels.

boiler: Closed container in which a liquid may be heated and vaporized.

boiling point: Temperature at which a liquid boils or generates bubbles of vapor when heated.

bond: A mechanical connection between metallic parts of an electrical system, with a bonding locknut or bushing, such as between a neutral wire and a meter enclosure or service conduit to the service-equipment enclosure; the junction of welded parts; the adhesive for abrasive grains in grinding wheels.

bonding bushing: A special conduit bushing equipped with a conductor terminal to take a bonding jumper; also has a screw or other sharp device to bite into the enclosure wall to bond the conduit to the enclosure without a jumper when there are no concentric knockouts left in the wall of the enclosure. See Fig. B-4.

bonding jumper: A bare or insulated conductor used to ensure the required electrical conductivity between metal parts required to be electrically connected. Frequently used from a bonding bushing to the service equipment enclosure to provide a path around concentric knockouts in an enclosure wall; also used to bond one raceway to another.

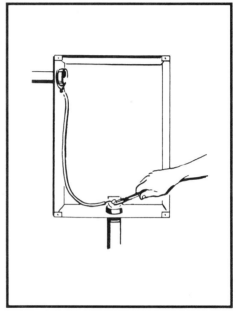

Fig. B-4: Panelboard housing ("can") utilizing bonding locknuts and bonding jumpers.

bonding locknut: A threaded locknut for use on the end of a conduit terminal, but a locknut equipped with a screw through its lip. When the locknut is installed, the screw is tightened so its end bites into the wall of the enclosure close to the edge of the knockout.

booster: A generator connected in series with a circuit for the purpose of increasing the voltage of that circuit. Generally used in connection with a system where a storage battery carries part of the load. The booster increases the voltage to a point necessary for charging the battery.

braid: An interwoven cylindrical covering of fiber or wire.

branch circuit: That portion of a wiring system extending beyond the final overcurrent device protecting a circuit.

branch splice: The connection where a wire or conductor taps off from another wire or conductor.

braze: The joining together of two metal pieces, without melting them, using heat and diffusion of a jointing alloy of capillary thickness.

breadboard: Laboratory idiom for an experimental circuit.

breakdown: The abrupt change of resistance from high to low, allowing current flow; an initial rolling or drawing operation.

breaker strip: Thin strips of material placed between phase conductors and the grounding conductor in flat parallel portable cables; the breaker strips provide extra mechanical and electrical protection.

breakout: The point at which conductor(s) are taken out of a multiconductor assembly.

bridge: A circuit that measures by balancing four impedances through which the same current flows:

- Wheatstone—resistance

- Kelvin—low resistance

- Schering—capacitance, dissipation factor, dielectric constant

- Wien—capacitance, dissipation factor

British thermal unit (Btu): Quantity of heat required to raise the temperature of 1 pound of water 1 degree Fahrenheit.

brush: A conductor between the stationary and rotating parts of a machine, usually of carbon. See F. B-5 on page 22.

brush holders: Adjustable arms for holding the commutator brushes of a generator against the commutator, feeding them forward to maintain proper contact as they wear, and permitting them to be lifted from contact when necessary.

brush loss: The loss in watts due to the resistance of the brush contact against the surface of the commutator.

buck: Rough wood door frame placed on a wall partition to which the door moldings are attached; completely fabricated steel door frame set in a wall or partition to receive the door.

buff: To lightly abrade.

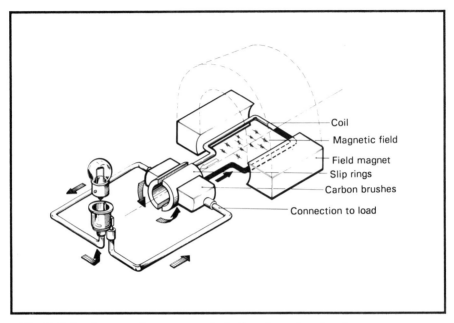

Fig. B-5: Basic parts of a generator including the carbon brushes.

bug: A crimped or bolted type of electrical connector for splicing wires or cables together. Also used as a verb: "bugged." Example: The wires were bugged together.

bull-cutters: A large, long-handled tool for cutting the larger sizes of wire and cable, up to kcmil sizes. See Fig. B-6.

bus: The conductor(s) serving as a common connection for two or more circuits. The neutral bus in a circuit breaker panel is shown in Fig. B-7. All branch-circuit neutral conductors terminate here. All ungrounded conductors terminate at the "hot" bus bars.

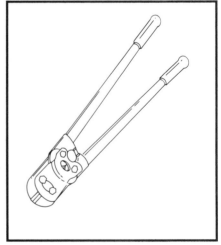

Fig. B-6: Bull-cutters are used for cutting the larger sizes of wire and cable.

busbars: The conductive bars used as the main current supplying elements of panelboards or switchboards; also the conductive bars duct: an assembly of bus bars within an enclosure that is designed for ease of installation, has no fixed electrical characteristics, and allows power to be taken off conveniently, usually without circuit interruption.

bushing: A metal, fiber, or plastic fitting designed to screw onto the ends of conduit and cable connectors to provide protection for conductors. Some metal bushings have provisions for bonding jumpers as shown in Fig. B-4 on page 20.

butane: Liquid hydrocarbon commonly used as fuel for heating purposes.

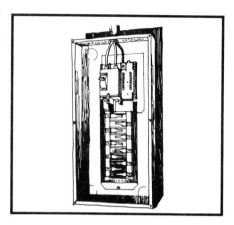

Fig. B-7: Circuit breaker panelboard showing bus bars, protected by a main circuit breaker, for ungrounded conductors and a neutral bus for grounded and bonding conductors.

buttress: Projecting structure built against a wall to give it greater strength.

butt welding: A weld in which the two pieces to be connected do not overlap but are welded directly at their ends; a common method of welding rods by an electric process.

buzzer: An electric call signal that makes a buzzing noise caused by the rapid vibrations of the armature. It operates on the same principle as the vibrating bell.

BWR (boiling water reactor-nuclear power): A basic nuclear power fission reactor in which steam is used to transfer the energy from the reactor.

BX: A nickname for type AC armored cable (wires with a spiral-wound, flexible steel outer jacketing); although used generically, BX is a registered trade-name of the General Electric Company. See Fig. B-8 on page 24.

bypass: Passage at one side of or around a regular passage. A bypass switch, for example, would route current around a device or machine.

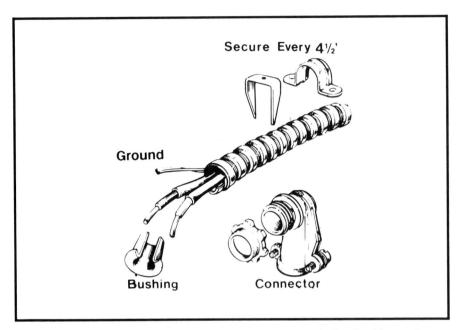

Fig. B-8: Type AC (armored or "BX") cable with insulating bushing and approved fasteners.

C

cable: An assembly of two or more wires that may be insulated or bare.

cable, aerial: An assembly of one or more conductors and a supporting messenger. Sometimes referred to as "messenger cable." See Fig. C-1.

cable, armored: A cable having armor (see armor).

cable, belted: A multiconductor cable having a layer of insulation over the assembled insulated conductors.

cable, bore-hole: The term given vertical-riser cables in mines.

cable clamp: A device used to clamp around a cable to transmit mechanical strain to all elements of the cable.

cable, coaxial: A cable used for high frequency, consisting of two cylindrical conductors with a common axis separated by a dielectric; normally the outer conductor is operated at ground potential for shielding.

cable, control: Used to supply voltage (usually ON or OFF) to motor controls and other controllers that operate electrical circuits or apparatus. Often used in cable trays.

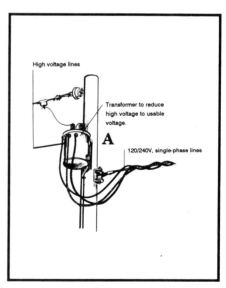

Fig. C-1: Aerial cable used for outside distribution.

cable, duplex: A twisted pair of cables.

cable, festoon: A cable draped in accordion fashion from sliding or rolling hangers, usually used to feed moving equipment such as bridge cranes.

cable, hand: A mining cable used to connect equipment to a reel truck.

cable, parkway: Designed for direct burial with heavy mechanical protection of jute, lead, and steel wires.

cable, portable: Used to transmit power to mobile equipment.

cable, power: Used to supply current (power).

cable, pressure: A cable having a pressurized fluid (gas or oil) as part of the insulation; paper and oil are the most common fluids.

cable, ribbon: A flat multiconductor cable.

cable, service drop: The cable from the utility line to the customer's property. See Fig. C-2.

cable, signal: Used to transmit data.

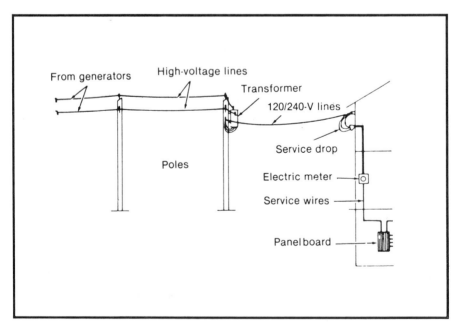

Fig. C-2: Typical overhead electric service drop supplying a residence.

cable, spacer: An aerial distribution cable made of covered conductors held by insulated spacers; designed for wooded areas.

cable, spread room: A room adjacent to a control room to facilitate routing of cables in trays away from the control panels.

cable, submarine: Designed for crossing under navigable bodies of water; hav-ing heavy mechanical protection against anchors, floating debris, and moisture.

cable, tray: A multiconductor having a nonmetallic jacket, designed for use in cable trays; (not to be confused with type TC cable for which the jacket must also be flame retardant).

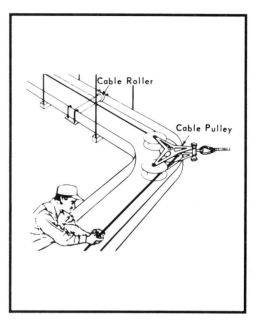

Fig. C-3: Practical application of a cable tray system.

cable tray: A rigid structure to support cables; a type of raceway; normally having the appearance of a ladder and open at the top to facilitate changes. See Fig. C-3.

cable, triplexed d: Helical assembly of 3 insulated conductors and sometimes a bare grounding conductor.

cable, unit: A cable having pairs of cables stranded into groups (units) of a given quantity, then these groups form the core.

cable, vertical riser: Cables utilized in circuits of considerable elevation change; usually incorporate additional components for tensile strength.

cabling: Helically wrapping together of two or more insulated wires.

caisson: Sunken panel in a ceiling, contributing to a pattern.

calender: A machine that mixes and makes slabs of polymers by squeezing heated, viscous material between two counter-rotating rollers.

calibrate: Compare with a standard.

calorie: Heat required to raise temperature of 1 gram of water 1 degree centigrade.

calorimeter: 1) An instrument for measuring the heat generated by an electrical current in a conductor. 2) An apparatus for measuring the quantity of heat generated by friction, combustion, or other chemical change.

cambric: A fine weave of linen or cotton cloth used as insulation.

candela (cd): The basic SI unit for luminous intensity: the candela is defined as the luminous intensity of $\frac{1}{600,000}$ of a square meter of a blackbody at the temperature of freezing platinum.

candlefoot: A unit of illumination. The light given by a British standard candle at a distance of one foot. See *footcandle.*

candle power: The illuminating power of a standard sperm candle used as a measure for other illuminants.

canopy switch: A small switch usually located in the canopy of an electric fixture.

cantilever: Projecting beam or member supported at only one end.

cant strip: Beveled strip placed in the angle between the roof and an abutting wall to avoid a sharp bend in the roofing material; strip placed under the lowest row of tiles on a roof to give it the same slope as the rows above it.

capacitance: The storage of electricity in a capacitor; the opposition to voltage change; the unit of measurement is the farad (f) or microfarad (mf).

capacitive reactance: The measure of resistance to the flow of an alternating current through a resistor.

capacitor: An apparatus consisting of two conducting surfaces separated by an insulating material. It stores energy, blocks the flow of direct current, and permits the flow of alternating current to a degree depending on the capacitance and frequency.

capillary action: The traveling of liquid along a small interstice due to surface tension.

capstan: A rotating drum used to pull cables or ropes by friction; the cables are wrapped around the drum.

carbon black: A black pigment produced by the incomplete burning of natural gas or oil; used as a filler.

carbon dioxide: Compound of carbon and oxygen that is sometimes used as a refrigerant.

carbonizing: The reduction of a substance to carbon by subjecting it to intense heat in a closed vessel.

carpet strip: A strip attached to the floor beneath a door.

carriage: The timber or steel joist that supports the steps of a wooden stair.

carrying capacity: The greatest amount of electrical current that a conductor can safely carry, expressed in amperes. The various size wires, with their carrying capacities, are arranged in tables in the National Electrical Code. See *current-carrying capacity.*

cartridge fuse: A fuse enclosed in an insulating tube to confine the arc when the fuse blows.

cascade: The output of one device connected to the input of another.

casing: The framework about a window or door.

catepuller: Two endless belts that squeeze and pull a cable by friction.

cathode: 1) The negative electrode through which current leaves a nonmetallic conductor, such as an electrolytic cell. 2) The positive pole of a storage battery. 3) Vacuum tube — the electrode that emits electrons.

cathode rays: Streams of electrons emitted from the filament (called the cathode) of a vacuum tube under the influence of high voltage and which, by suitable means, can be brought outside the tube.

cathode-ray tube: The electronic tube which has a screen upon which a beam of electrons from the cathode can be made to create images; for example, the television picture tube.

cathodic protection: Reduction or prevention of corrosion by making the metal to be protected the cathode in a direct-current circuit.

cation: The element or positive ion that appears at the cathode or negative terminal in an electrolytic cell.

caulking: Making a joint or seam watertight or steam-tight by filling it in with rust cement or by closing the joint by means of a caulking tool. Oakum is frequently caulked into the seams of wooden vessels.

cavity wall: Wall built of solid masonry units arranged to provide airspace within the wall.

CB: Pronounced "see bee," an expression used to refer to "circuit breaker," taken from the initial letters C and B.

C-C: Center to center.

CCA (Customer Complaint Analysis): A formal investigation of a cable defect or failure.

CEE (International Commission on Rules for the Approval of Electrical Equipment): Controls the standards for electrical products for sale in Europe; analogous to UL in USA.

cell: A single element of an electric battery, either primary or secondary.

Centigrade scale: Temperature scale used in metric system. Freezing point of water is 0 degrees; boiling point is 100 degrees.

centrifugal switch: A switch used on the single-phase, split-phase motor to open the starting winding after the motor has almost reached synchronous speed.

CFR (Code of Federal Regulations): The general and permanent rules published in the Federal Register by the executive departments and agencies of the Federal Government. The Code is divided into 50 Titles that represent broad areas; Titles are divided into Chapters that usually bear the name of the issuing agency, e.g., Title 30 — Mineral Resources, Chapter I = *MESA*; Title 29 — Labor, Chapter XVII — *OSHA*; Title 10 — Energy, Chapter I = *NRC*.

CFR Title 10, Chapter I (nuclear power): The regulations of the Federal Nuclear Regulatory Commission: a) Standards for protection against radiation. b) Licensing procedures of production and utilization facilities. c) Operators' licenses. d) Special nuclear materials. e) Nuclear material packaging for transport. f) Reactor site criteria.

cgs: Centimeter, gram, second.

chamfer: A beveled edge or cut-off corner.

charge: The quantity of positive or negative ions in or on an object; the quantity of electricity residing on an electrostatically charged body. Unit: coulomb.

charged cell: A storage cell that has had direct current passed through it until the positive plate has changed chemically fro $PbSO_4$ to $PbSO_2$ and the negative plate has changed from $PbSO_4$ to Pb.

charging current: Direct current applied to a storage battery to produce chemical action to charge the battery. Its direction is always the reverse of the discharge current.

charging rate: The rate of flow, in amperes, of electric current flowing through a storage battery while it is being charged.

chase: Recess in inner face of masonry wall providing space for pipes and/or ducts.

check valve: A device that permits fluid flow only in one direction.

chemical resistance test: Checking performance of materials immersed in different chemicals; loss of strength and dimensional change are measured.

chimney effect: Tendency of air or gas to rise when heated.

choke coil: A coil used to limit the flow of alternating current while permitting direct current to pass.

circuit: A closed path through which current flows from a generator, through various components, and back to the generator.

circuit breaker: A resettable fuse-like device designed to protect a circuit against overloading. See Fig. C-4.

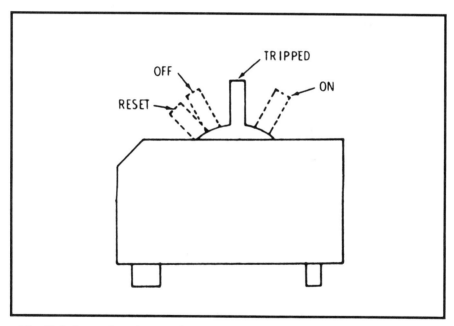

Fig. C-4: Operating characteristics of a typical one-pole circuit breaker.

circuit foot: One foot of circuit; i.e., if one has a 3-conductor circuit, then each lineal foot of circuit would have 3 circuit feet.

circular mil: The non-SI unit for measuring the cross-sectional area of a conductor.

CL: Center line; sometimes written C$_L$.

Class 1E (nuclear power): The safety electrical systems that are essential to emergency reactor shutdown, cooling, and containment.

Class 2 (nuclear power): Items important to reactor operation but not essential to safe shutdown or isolation.

clearance: The vertical space between a cable and its conduit.

clearing time: The time from sensing an overcurrent to circuit interruption.

closing die: A die used to position the individual conductors during cabling.

coated wire: Wire given a thin coating of another metal such as tin, lead, nickel, etc.; coating by dipping or planting; coating for protection, or to improve its properties.

coaxial cable: A cable consisting of two conductors concentric with and insulated from each other.

code: Any set of rules devised for the purpose of securing uniformity in work and for the maintaining of proper standards is usually called a code; that is, National Electrical Code, etc.

code installation: An installation that conforms to the local code and/or the national code for safe and efficient installation.

coefficient of expansion: The change in dimension due to change in temperature.

coefficient of friction: The ratio of the tangential force needed to start or maintain uniform relative motion between two contacting surfaces to the perpendicular force holding them in contact.

coil: A wire or cable wound in a series of closed loops; successive turns of insulated wire that create a magnetic field when an electric current passes through them.

cold bend: A test to determine cable or wire characteristics at low temperatures.

cold cathode: A cathode that does not depend on heat for electron emission.

cold joint: Improper solder connection due to insufficient heat.

cold welding: Solid-phase welding using pressure without heat.

cold work: Permanent strain produced by an external force (such as wire drawing) in a metal below its recrystallization temperature.

collar beam: A beam above the lower ends of the rafters and attached to them.

collector: The part of a transistor that collects electrons.

color code: Identifying conductors by the use of color.

combustion: The chemical union of a combustible substance with oxygen, resulting in the production of heat.

come along: A cable grip (usually of tubular basketweave construction which tightens its grip on the cable as it is pulled) with a pulling "eye" on one end for attaching to a pull-rope for pulling conductors into conduit or other raceway.

comfort zone: Area on psychrometric chart that shows conditions of temperature, humidity, and sometimes air movement in which most people are comfortable.

common failure mode (nuclear power): An event causing redundant equipment to be inoperable.

commutating pole: An electromagnetic bar inserted between the pole pieces of a generator to offset the cross magnetization of the armature currents.

commutator: Device used on electric motors or generators to maintain a unidirectional current.

Compax® die: A wire drawing die made by *GE* of sintered diamond.

compensated wattmeter: A wattmeter that has a compensating coil connected in series with the potential coil for the purpose of correcting the error caused by the absorption of power due to mechanical operation of the potential coil.

compensating coil: A coil that serves to compensate for the mechanical friction in the moving coil of a meter.

completed circuit: Also called closed circuit. One that has been made or closed.

compound fill: An insulation that is poured into place while hot.

compound wound: A generator or motor having a part of a series-field winding wound on top of a part of a shunt-field winding on each of the main pole pieces.

compressibility: A density ratio determined under finite testing conditions.

compression lug or splice: Installed by compressing the connector onto the strand, hopefully into a cold weld.

compressor: The pump of a refrigerating mechanism that draws a vacuum or low pressure on the cooling side of a refrigerant cycle and squeezes or compresses the gas into the high pressure or condensing side of the cycle.

computer: An electronic apparatus: 1) For rapidly solving complex and involved problems, usually mathematical or logical. 2) For storing large amounts of data.

concealed: Rendered inaccessible by the structure or finish of the building. Wires in concealed raceways are considered concealed, even though they may become accessible by withdrawing them.

concentricity: The measurement of the center of the conductor with respect to the center of the insulation.

condenser: The accumulator of electrical energy, as in a capacitor. A vessel in which the condensation of gases is effected.

conductance: The ability of material to carry an electric current; that is, the ease with which a conductor carries an electric current. The opposite of resistance.

conduction: The flow of an electric current through a conducting body, such as a copper wire.

conductivity: The relative ability of materials to carry an electrical current.

conductor: Any substance that allows energy flow through it with the transfer being made by physical contact but excluding net mass flow.

conductor, bare: Having no covering or insulation whatsoever.

conductor, covered: A conductor having one or more layers of nonconducting materials that are not recognized as insulation under the National Electric Code.

conductor, insulated: A conductor covered with material recognized as insulation.

conductor load: The mechanical loads on an aerial conductor—wind, weight, ice, etc.

conductor, plain: Consisting of only one metal.

conductor, segmental: Having sections isolated, one from the other and connected in parallel; used to reduce ac resistance.

conductor, solid: A single wire.

conductor, stranded: Assembly of several wires, usually twisted or braided.

conductor stress control: The conducting layer applied to make the conductor a smooth surface in intimate contact with the insulation; formerly called extruded strand shield (ESS).

conduit: A tubular raceway such as electrical metallic tubing (EMT); rigid metal conduit, rigid nonmetallic conduit, etc.

conduit fill: Amount of cross-sectional area used in a raceway.

conduit, rigid metal: Conduit made of Schedule 40 pipe, normally 10 foot lengths.

Condulet: A trade name for conduit fitting.

configuration, cradled: The geometric pattern that cables will take in a conduit when the cables are pulled in parallel and the ratio of the conduit ID to the 1/C cable OD is greater than 3.0.

configuration, triangular: The geometric pattern that cables will take in a conduit when the cables are triplexed or are pulled in parallel with the ratio of the conduit ID to the 1/C cable OD less than 2.5.

connection: That part of a circuit that has negligible impedance and joins components or devices.

connection, delta: Interconnection of 3 electrical-equipment windings in delta (triangular) fashion.

connection (nuclear power): A cable terminal, splice, or seal at the interface of the cable and equipment.

connection, star: Interconnection of 3 electrical-equipment windings in star (wye) fashion.

connector: A device used to physically and electrically connect two or more conductors. See Fig. C-5.

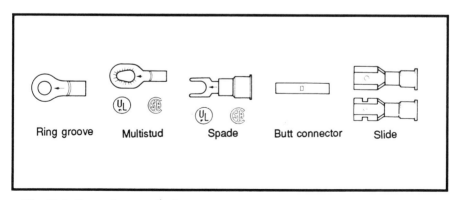

Fig. C-5: Several types of wire connectors.

connector, pressure: A connector applied using pressure to form a cold weld between the conductor and the connector.

connector, reducing: Used to join two different size conductors.

constant current: A type of power system in which the same amount of current flows through each utilization equipment, used for simplicity in street lighting circuits.

constant voltage: The common type of power in which all loads are connected in parallel with different amounts of current flowing through each load.

contact: A device designed for repetitive connections.

contactor: A type of relay.

containment (nuclear power): The safety barrier designed to prevent the release of radioactive material in case of reactor accident.

continuity: The state of being whole, unbroken.

continuous beam: A beam that rests on more than two supports.

continuous duty: That demand on a system requiring operation at a constant load for an indefinite period of time.

continuous load: 1) NEC—in operation three hours or more. 2) Nuclear power—8760 hours/year (scheduled maintenance outages permitted).

continuous vulcanization (CV): A system utilizing heat and pressure to vulcanize insulation after extrusion onto wire or cable; the curing tube may be in a horizontal or vertical pole.

control: Automatic or manual device used to stop, start, and/or regulate flow of gas, liquid, and/or electricity.

controller: A device or group of devices that serves to govern in some predetermined manner the electric power delivered to the apparatus to which it is connected.

control, temperature: A thermostatic device that automatically stops and starts a motor, the operation of which is based on temperature changes.

convection: The transfer of heat to a fluid by conduction as the fluid moves past the heat source.

convenience outlet: A point on the wiring system at which current is taken to supply portable 120-volt appliances such as tv sets, toasters, etc.

cook annealer: An annealer using heavy electrical current through the conductor as the heat source.

cook buncher: A buncher using controlled diameter and wire position.

cooling tower: Device that cools water by water evaporation in air. Water is cooled to the wet bulb temperature of air.

coordination: The selection of system components to prevent the failure of the whole system due to cascading; limiting system failure by activation of the fewest overcurrent devices, hopefully to one.

co-polymer: A polymer having two "repeating units."

copper: A word used by itself to refer to copper conductors. Examples: "A circuit of 500 MCM (kcmil) copper" or "the copper cost of the circuit." It is a good conductor of electricity, easily formed, easily connected to itself and other metals used in electrical wiring.

Copper Development Association: A manufacturer's group to promote the use of copper.

copper, electrolytic: Copper of high purity, refined by electrolysis, used for electrical conductors.

copper loss: The energy dissipated in the copper conductors of a circuit, due to heat loss of I^2R produced by current flow through the conductor. Term is sometimes used to refer to the same type of loss in aluminum circuit conductors.

Copperweld®: This is the trade name for a conductor composed of a steel core with a heavy copper coating. It is used where strength is needed such as large, long, vertical risers; most commonly used for ground rods.

cord: A small flexible conductor assembly, usually jacketed.

cord set: A cord having a wiring connector on one or more ends.

core: The portion of a foundry mold that shapes the interior of a hollow casting.

core (cable): The portion of an insulated cable under a protective covering.

cored hole: At the time of casting, a sand core is placed in the mold so that the metal flows around it. When the casting is cold, the sand core is broken away, leaving the hole.

core disks: Thin circular disks of sheet iron used for building laminated armature cores.

core loss: The electric loss occurring in the core of an armature or transformer due to eddy currents, hysteresis, and like influences.

corona: A low energy electrical discharge caused by ionization of a gas by an electric field.

corrosion: The deterioration of a substance (usually a metal) because of a reaction with its environment.

coulomb: A unit of electrical quantity. The derived SI unit for quantity of electricity or electrical charge: one coulomb equals one ampere-second.

counterbore: A tool that enlarges an already-machined round hole to a certain depth. The pilot of the tool fits in the smaller hole, and the larger part counterbores or makes the end of the hole larger.

counter emf: The voltage opposing the applied voltage and the current in a coil; caused by a flow of current in the coil; also known as back emf; induced voltage.

coupled-inductance: Voltage induced in one circuit by current changes in a second circuit.

coupling: The means by which signals are transferred from one circuit to another. The mechanical connector between two pieces of conduit.

coupon: A piece of metal for testing, of specified size; a piece of metal from which a test specimen may be prepared.

CPE (chlorinated polyethylene): A plastic for cable jackets.

cramp: Iron rod with ends bent to a right angle; used to hold blocks of stone together.

crawl space: Shallow space between the first tier of beams and the ground (no basement).

crazing: Fine cracks that may extend in a network on or under the surface of a material; usually occurs in the presence of an organic liquid or vapor.

cross head: The mechanism on an extruder where the material is applied; it holds the die, guider, and core tube; usually just called "head."

cross-link: To cure by linking molecules together in a polymer—either by using chemical cross-linking agents or radiation.

cross-linked polyethylene: Thermosetting polyethylene that has better physical properties than plain polyethylene: used as an insulation having good physical properties.

cross talk: Undesired pickup of signals by a second circuit.

CRT: Cathode ray tube.

crystal: A solid composed of atoms, ions, or molecules arranged in a pattern that is periodic in three dimensions.

CT: Pronounced "see tee"; refers to current transformer, taken from the initial letters C and T.

CU: Copper.

cure: To change the properties of a polymeric system into a more stable, usable condition by the use of heat, radiation, or reaction with chemical additives.

current (I): The time rate of flow of electric charges; unit: ampere.

current-carrying capacity: The current in amperes a conductor can carry continuously under the conditions of use without exceeding its temperature rating.

current, charging: The current needed to bring the cable up to voltage; determined by capacitance of the cable; after withdrawal of voltage, the charging current returns to the circuit; the charging current will be 90° out of phase with the voltage.

current density: The current per unit cross-sectional area.

current-induced: Current in a conductor due to the application of a time-varying electromagnetic field.

current, leakage: That small amount of current that flows through insulation whenever a voltage is present and heats the insulation because of the insulation's resistance; the leakage current is in phase with the voltage, and is a power loss.

current limiting: A characteristic of short-circuit protective devices, such as fuses, by which the device operates so fast on high short circuit currents that less than a quarter wave of the alternating cycle is permitted to flow before the circuit is opened, thereby limiting the thermal and magnetic energy to a certain maximum value, regardless of the current available.

curtain wall: A thin wall supported independently of the wall below, by the structural steel or concrete frame of the building.

cut-in: The connection of electrical service to a building, from the power company line to the service equipment, e.g., "the building was cut-in" or "the power company cut-in the service."

cut-in-card: The certificate of approval issued by the electrical inspection authority to the electrical contractor, to be given to the power company as evidence that the building electrical system is safe for connection or "cut-in" by the power company.

cut nails: Machine-cut iron nails as distinguished from wire nails.

cutout: A fuse holder that may be used to isolate part of a circuit.

cutout box: A surface mounting enclosure with a cover equipped with swinging doors, used to enclose fuses.

cutover: Changing from one reel to another without stopping the manufacturing process.

cut resistance: The ability of a material to withstand mechanical pressure without rupture or becoming ineffective.

cycle: 1) An interval of space or time in which one set of events or phenomena is completed. 2) A set of operations that are repeated regularly in the same sequence. 3) When a system in a given state goes through a number of different processes and finally returns to its initial state.

cyclic aging: A test on a closed loop of cable having voltage applied, and induced current applied in cycles to cause expansion and contraction; simulates cable operating in a dry environment.

D

damper: Valve for controlling air flow.

damping: The dissipation of energy with time or distance.

damping coil: A coil mounted near a galvanometer to produce a damping effect; i.e., to bring the needle quickly to a point of rest after deflection.

Daniell cell: A closed-circuit type of primary cell.

D'Arsonval galvanometer: A very sensitive periodic or dead-beat galvanometer in which the indicating coil is suspended in the field of a powerful horseshoe magnet.

DBE (Design Basis Event) (nuclear power): Postulated abnormal events used to establish the performance requirements of the structures, systems, and components.

dead: 1) Not having electrical charge. 2) Not having voltage applied.

dead beat: Instruments where indicators come promptly to a position of rest due to heavy damping.

dead-end: A mechanical terminating device on a building or pole to provide support at the end of an overhead electric circuit. A dead-end is also the term used to refer to the last pole in the pole line. The pole at which the electric circuiting is brought down the pole to go underground or to the building served.

dead-front: A switchboard or panel or other electrical apparatus without "live" energized terminals or parts exposed on front where personnel might make contact.

dead load: A load whose pressure is steady and constant.

deadman: Reinforced concrete anchor set in earth and tied to the retaining wall for stability.

deadman's switch: A switch necessitating a positive action by the operator to keep the system or equipment running or energized.

debug: To examine or test a procedure, routine, or equipment for the purpose of detecting and correcting errors, especially during start-up.

decay (nuclear power): The transmutation of a nucleus to a stable energy condition.

decibel: A unit for measuring sound intensity, named in honor of Alexander Graham Bell. When sound or noise is created energy is given off which is measured in decibels; i.e., the noise of an airplane engine measures 120 decibels.

deeping: Cutting out to a depth; placing comparatively far below the surrounding surface.

defeater: A means to deactivate a safety interlock system.

defense in depth (nuclear power): A basic design philosophy to keep nuclear power plants safe during normal operations and the worst imagined accidents. There are 3 levels of defense: 1) Accident prevention, quality assurance, redundancy, inspection testing. 2) Protection devices and systems. 3) Safety systems to function in case 1 and 2 fail.

deflection: Deviation of the central axis of a beam from normal when the beam is loaded.

deflection plate: The part of a certain type of electron tube that provides an electrical field to produce deflection of an electron beam.

delta connection: The connection of the circuits in a three-phase system in which the terminal connections are triangular like the Greek letter delta.

demagnetization: The process of removing magnetism from a magnetized substance.

demand: 1) The measure of the maximum load of a utility's customer over a short period of time. 2) The load integrated over a specified time interval.

demand factor: For an electrical system or feeder circuit, this is a ratio of the amount of connected load (in kVA or amperes) that will be operating at the same time to the total amount of connected load on the circuit. An 80% demand factor, for instance, indicates that only 80% of the connected load on a circuit will ever be operating at the same time. Conductor capacity can be based on that amount of load.

demonstration (nuclear power): A course of reasoning showing that a certain result is a consequence of assumed premises; an explanation or illustration.

density: Closeness of texture or consistency.

depolarization: The process of preserving the activity of a voltaic cell by preventing polarization.

depolarizer: An oxidizing substance used for fixing the hydrogen derived from the decomposition of the acid by the zinc in primary cells.

derating: The intentional reduction of stress/strength ratio in the application of a material; usually for the purpose of reducing the occurrence of a stress-related failure.

derating factor: A factor used to reduce ampacity when the cable is used in environments other than the standard, or when more than the standard number of conductors is installed in a raceway.

detection: The process of obtaining the separation of the modulation component from the received signal.

deviation: Departure from the exact or from a set standard.

device: An item intended to carry or help carry, but not utilize, electrical energy. See Fig. D-1.

dew point: The temperature at which vapor starts to condense (liquify) from a gas-vapor mixture at constant pressure.

dial: A graduated plate, usually circular or oval, on which a reading is indicated by a needle or pointer.

die: 1) Wire: a metal device having a conical hole that is used to reduce the diameter of wire drawn through the die or series of dies. 2) Extruder: the fixed part of the mold. 3) An internal screw used for cutting an outside thread.

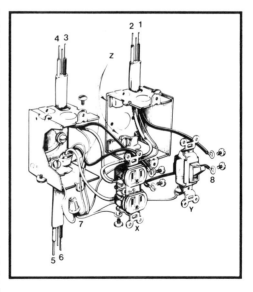

Fig. D-1: The single-pole switch and duplex receptacle are known as wiring devices since they do not actually use electricity for operation.

dielectric: An insulator or a term referring to the insulation between the plates of a capacitor.

dielectric absorption: The storage of charges within an insulation; evidenced by the decrease of current flow after the application of dc voltage.

dielectric constant: The ratio of the conductivity of a dielectric for electrostatic lines of force to that of air.

dielectric dispersion: The change in relative capacitance due to change in frequency.

dielectric heating: The heating of an insulating material by ac induced internal losses; normally frequencies above 10 mHz are used.

dielectric loss: The time rate at which electrical energy is transformed into heat in a dielectric when it is subjected to a changing electric field.

dielectric phase angle: The phase angle between the sinusoidal ac voltage applied to a dielectric and the component of the current having the same period.

dielectric strength: The maximum voltage that an insulation can withstand without breaking down; usually expressed as a gradient — vpm (volts per mil).

differential motor: A motor with a compound-wound field, in which the series and shunt coils oppose each other.

dimension: A definite measurement shown on a drawing, as length, width, or thickness. Dimensions should be given on all working drawings, whether drawn to scale or not, but not on assembly drawings, except "overall" dimensions.

dimensioning: Indicating on a drawing the sizes of various parts.

dimension line: A line on a drawing that indicates to what part or line the dimension has reference.

diode: A device having two electrodes: the cathode and the plate or anode — used as a rectifier and detector. See Fig. D-2.

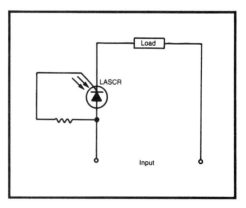

Fig. D-2: Schematic drawing of a diode.

direct current (dc): Electricity that flows only in one direction; produced by a battery and dc generators.

direct drive: A compact arrangement of driving a generator by direct connection with the prime mover or the load, avoiding the use of shafts and belts.

direction of lay: The lateral direction, designated as left-hand or right-hand, in which the elements of a cable run over the top of the cable as they recede from an observer looking along the axis of the cable.

direction of magnetic flux: The direction is always from the north to the south pole in a magnetic field.

disconnect: A switch for disconnecting an electrical circuit or load (motor, transformer, panel) from the conductors that supply power to it, e. g., "He pulled the motor disconnect," meaning he opened the disconnect switch to the motor.

disconnecting means: A device, a group of devices, or other means whereby the conductors of a circuit can be disconnected from their supply source.

dispersion: Holding fine particles in suspension throughout a second substance.

displacement current: An expression for the effective current flow across a capacitor.

dissipation factor: Energy lost when voltage is applied across an insulation because of capacitance: the cotangent of the phase angle between voltage and current in a reactive component; because the shift is so great, we use the complement (angle) of the angle ϕ which is used for power factor; dissipation factor = tan = cot ϕ: is quite sensitive to contamination and deterioration of insulation: also known as power factor (of dielectrics).

distortion: Unfaithful reproduction of signals.

distribution box: A large metal box used in conduit installation as a center of distribution.

distribution line: An exterior supply line from which individual installations are supplied.

distribution, statistical analysis: A statistical method used to analyze data by correlating data to a theoretical curve to a) Test validity of data. b) Predict performance at conditions different from those used to produce the data: The normal distribution curve is most common.

diversity factor: The ratio of the sum of load demands to a system demand. See *demand factor*.

DOAL: Diameter Overall.

DOC: Diameter Over Conductor; note that for cables having a stress control, the diameter over the stress control layer becomes DOC.

documents (nuclear power): Pertaining to Class 1E equipment and systems.

DOI: Diameter Over Insulation.

DOIS: Diameter Over Insulation Shield.

DOJ: Diameter Over Jacket.

donkey: A motor-driven power machine on legs or wheels used for threading and/or cutting conduit. Sometimes called "mule."

doorframe: The surrounding case into and out of which a door opens and shuts.

DOSC: Diameter Over Stress Control.

dose, radiation: The amount of energy per unit mass of material deposited at each point of an object undergoing radiation.

double-strength glass: One-eighth inch thick sheet glass (glass rated as single strength is $\frac{1}{10}$ inch thick).

dovetail: An interlocking joint.

draft indicator: An instrument used to indicate or measure chimney draft or combustion gas movement.

draft stop or fire stop: Obstructions placed in air passages to prevent the passage of flames up or across a building.

draintile: Hollow tile used for draining wet places.

drain wire: A bonding wire laid parallel to and touching shields.

drawing: Reducing wire diameter by pulling through dies.

drawing, block diagram: A simplified drawing of a system showing major items as simplified blocks; normally used to show how the system operates and the relationship between major items. See Fig. D-3.

drawing, electrical: Consists of lines, symbols, dimensions, and notations to accurately convey an engineer's design to workmen who install the electrical system on the job. A means of conveying a large amount of exact, detailed information in an abbreviated language.

drawing, line schematic (diagram): Shows how a circuit works. See Fig. D-4.

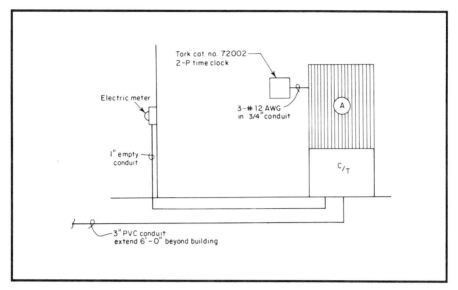

Fig. D-3: A power-riser (block) diagram used to show the arrangement of panels and feeders for an electrical system.

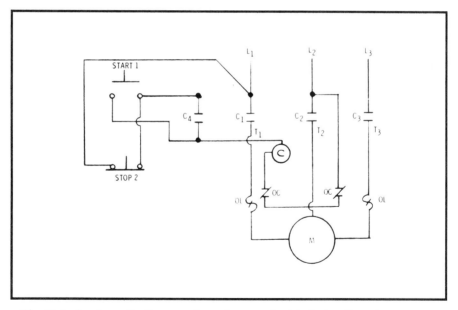

Fig. D-4: A schematic diagram shows how an electrical circuit works.

drawing, plot or layout: Shows the "floor plan." See Fig. D-5.

drawing, wiring diagram: Shows how electrical devices are interconnected. See Fig. D-6.

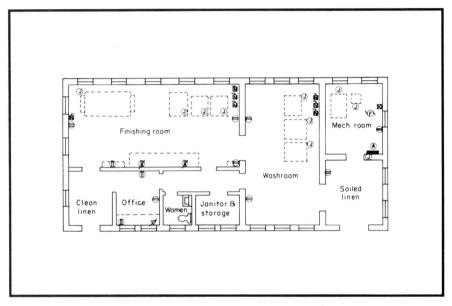

Fig. D-5: A floor plan of a building is drawn as if the roof was removed and the viewer is looking down at the building from above.

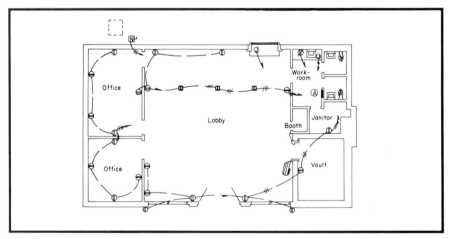

Fig. D-6: A wiring diagram shows how devices are connected in an electrical circuit.

drill: A circular tool used for machining a hole.

drip: Projecting horizontal course sloped outward to throw water away from a building.

drip loop: An intentional sag placed in service entrance conductors where they connect to the utility service drop conductors on overhead services; the drip loop will conduct any rain water to a point lower than the service head, to prevent moisture being forced into the service conductors by hydrostatic pressure and then running through the service head into the service conduit or cable. See Fig. D-7.

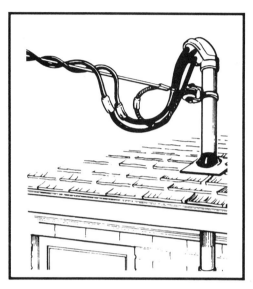

Fig. D-7: A drip loop conducts rain water to a point lower than the service head.

drum: The part of a cable reel on which the cable is wound.

drum armature: A generator or motor armature having its coils wound longitudinally or parallel to its axis.

dry: Not normally subjected to moisture.

dry battery: An electric battery made up of a number of dry voltaic cells. The term is often wrongly applied to a single dry cell.

dry bulb: An instrument with a sensitive element that measures ambient (moving) air temperature.

dry cell: A primary cell that does away with the liquid electrolyte so that it may be used in any position.

drywall: Interior wall construction consisting of plasterboard, wood paneling, or plywood nailed directly to the studs without application of plaster.

dual extrusion: Extruding two materials simultaneously using two extruders feeding a common cross head.

duct: A tube or channel through which air is conveyed or moved.

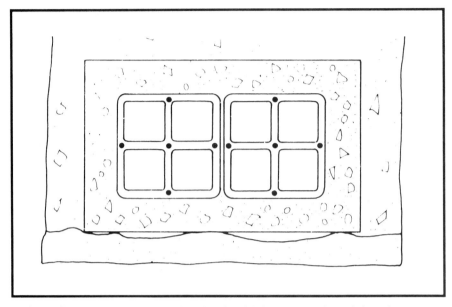

Fig. D-8: Cross-section of a duct bank utilizing grouped conduit to house conductors.

duct bank: Several underground conduits grouped together. See Fig. D-8.

ductility: The ability of a material to deform plastically before fracturing.

dumbbell: A die-cut specimen of uniform thickness used for testing tensile and elongation of materials.

durometer: An instrument to measure hardness of a rubber-like material.

duty, continuous: A service requirement that demands operation at a substantially constant load for an indefinitely long time.

duty, intermittent: A service requirement that demands operation for alternate intervals of load and no load, load and rest, or load, no load, and rest.

duty, periodic: A type of intermittent duty in which the load conditions regularly reoccur.

duty, short-time: A requirement of service that demands operations at loads and for intervals of time, both of which may be subject to wide variation.

dwarf partition: Partition that ends short of the ceiling.

dwell: A planned delay in a timed control program.

dwelling: A term of rather broad use, meaning a house or residence; the National Electrical Code recognizes a house as a single-family dwelling and an apartment as a multi-family dwelling. See Fig. D-9.

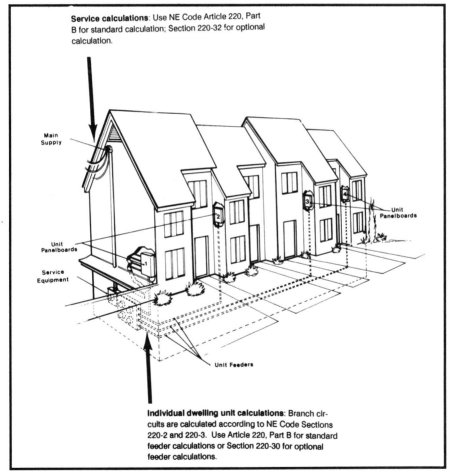

Figure D-9: An apartment complex is recognized by the NE Code as a multi-family dwelling unit.

dynamic: A state in which one or more quantities exhibit appreciable change within an arbitrarily short time interval.

dynamo: An electrical machine for converting mechanical energy into electrical energy; an electrical generator, especially for producing direct current. See Fig. D-10.

dynamometer: A device for measuring power output or power input of a mechanism.

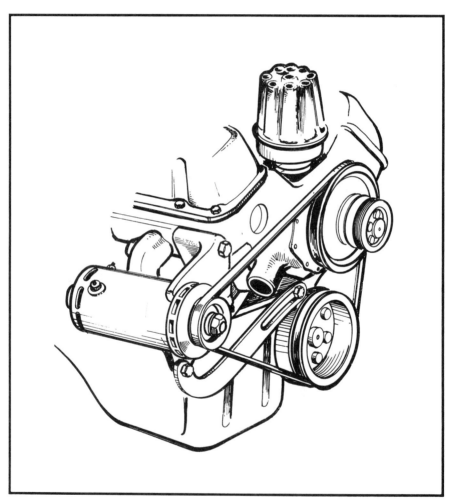

Fig. D-10: An automotive generator is one of the smallest forms of a dynamo.

E

E: Symbol for voltage, electrical pressure, electromotive force.

eaves: The projecting edges of a roof.

EC: Electrical Conductor of Aluminum.

eccentricity: 1) A measure of the entering of an item within a circular area. 2) The percentage ratio of the difference between the maximum and minimum thickness to the minimum thickness of an annular area.

ECCS (Emergency Core Cooling System) (nuclear power): A system to flood the fueled portion of the reactor and remove the residual heat produced by radioactive decay.

eddy currents: Circulating currents induced in conducting materials by varying magnetic fields; usually considered undesirable because they represent loss of energy and cause heating.

Edison base: The standard screw base used for ordinary lamps and Edison-base plug fuses.

EEI: Edison Electric Institute.

effective temperature: Overall effect on a person of air temperature, humidity, and air movement.

efficiency: The ratio of the output to the input.

elasticity: That property of recovering original size and shape after deformation.

elastomer: A material which, at room temperature, stretches under low stress to at least twice its length and snaps back to the original length upon release of stress.

elbow: A short conduit that is bent — usually at a 90-degree angle.

electrical symbols: Graphical symbols used on electrical drawings. In preparing electrical drawings, most engineers and designers used symbols adopted by the United States of America Standards Institute (USASI). However, many designers frequently modify these standard symbols to suit their own needs. For this reason, most electrical working drawings for building construction will have a symbol list or legend. The symbols shown in Fig. E-2 (beginning on page 55) are typical of those found on electrical drawings.

electric defrosting: Use of electric resistance heating coils to melt ice and frost off evaporators during defrosting.

electric heating: House heating system in which heat from electrical resistance units is used to heat rooms. See Fig. E-1.

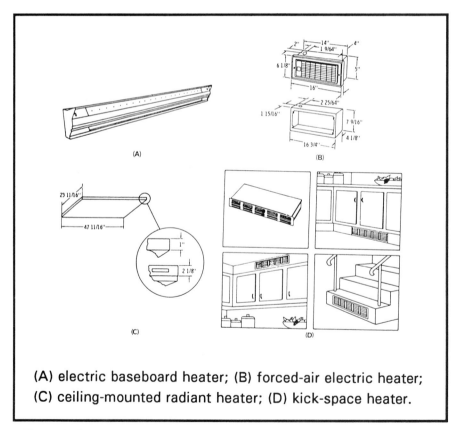

(A) electric baseboard heater; (B) forced-air electric heater; (C) ceiling-mounted radiant heater; (D) kick-space heater.

Fig. E-1: Several types of electric heating units used in residential applications.

Electrical Reference Symbols

ELECTRICAL ABBREVIATIONS (Apply only when adjacent to an electrical symbol.)	
Central Switch Panel	CSP
Dimmer Control Panel	DCP
Dust Tight	DT
Emergency Switch Panel	ESP
Empty	MT
Explosion Proof	EP
Grounded	G
Night Light	NL
Pull Chain	PC
Rain Tight	RT
Recessed	R
Transfer	XFER
Transformer	XFRMR
Vapor Tight	VT
Water Tight	WT
Weather Proof	WP

ELECTRICAL SYMBOLS

Switch Outlets

Single-Pole Switch	S
Double-Pole Switch	S_2
Three-Way Switch	S_3
Four-Way Switch	S_4
Key-Operated Switch	S_K

Switch and Fusestat Holder	S_{FH}
Switch and Pilot Lamp	S_P
Fan Switch	S_F
Switch for Low-Voltage Switching System	S_L
Master Switch for Low-Voltage Switching System	S_{LM}
Switch and Single Receptacle	⊖S
Switch and Duplex Receptacle	⊜S
Door Switch	S_D
Time Switch	S_T
Momentary Contact Switch	S_{MC}
Ceiling Pull Switch	Ⓢ
"Hand-Off-Auto" Control Switch	HOA
Multi-Speed Control Switch	M
Push Button	▪

Receptacle Outlets

Where weather proof, explosion proof, or other specific types of devices are to be required, use the upper-case subscript letters. For example, weather proof single or duplex receptacles would have the uppercase WP subscript letters noted alongside of the symbol. All outlets should be grounded.

Single Receptacle Outlet	⊖
Duplex Receptacle Outlet	⊜
Triplex Receptacle Outlet	⊕
Quadruplex Receptacle Outlet	⊕

Fig. E-2: Recommended electrical symbols.

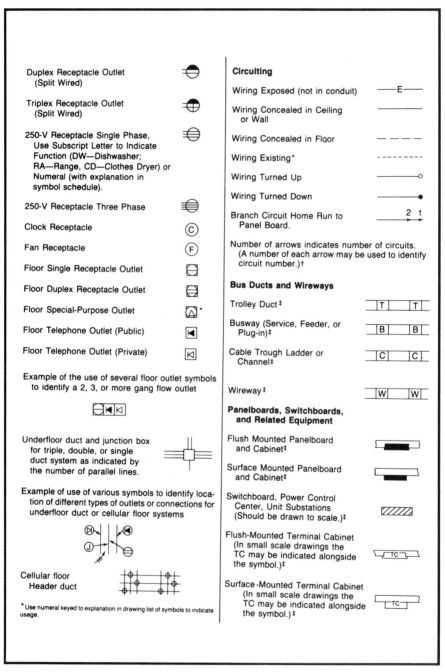

Duplex Receptacle Outlet
(Split Wired)

Triplex Receptacle Outlet
(Split Wired)

250-V Receptacle Single Phase,
Use Subscript Letter to Indicate
Function (DW—Dishwasher;
RA—Range, CD—Clothes Dryer) or
Numeral (with explanation in
symbol schedule).

250-V Receptacle Three Phase

Clock Receptacle

Fan Receptacle

Floor Single Receptacle Outlet

Floor Duplex Receptacle Outlet

Floor Special-Purpose Outlet

Floor Telephone Outlet (Public)

Floor Telephone Outlet (Private)

Example of the use of several floor outlet symbols
to identify a 2, 3, or more gang flow outlet

Underfloor duct and junction box
for triple, double, or single
duct system as indicated by
the number of parallel lines.

Example of use of various symbols to identify location of different types of outlets or connections for
underfloor duct or cellular floor systems

Cellular floor
Header duct

*Use numeral keyed to explanation in drawing list of symbols to indicate usage.

Circuiting

Wiring Exposed (not in conduit)

Wiring Concealed in Ceiling
or Wall

Wiring Concealed in Floor

Wiring Existing*

Wiring Turned Up

Wiring Turned Down

Branch Circuit Home Run to
Panel Board.

Number of arrows indicates number of circuits.
(A number of each arrow may be used to identify
circuit number.)†

Bus Ducts and Wireways

Trolley Duct‡

Busway (Service, Feeder, or
Plug-in)‡

Cable Trough Ladder or
Channel‡

Wireway‡

**Panelboards, Switchboards,
and Related Equipment**

Flush Mounted Panelboard
and Cabinet‡

Surface Mounted Panelboard
and Cabinet‡

Switchboard, Power Control
Center, Unit Substations
(Should be drawn to scale.)‡

Flush-Mounted Terminal Cabinet
(In small scale drawings the
TC may be indicated alongside
the symbol.)‡

Surface-Mounted Terminal Cabinet
(In small scale drawings the
TC may be indicated alongside
the symbol.)‡

Fig. E-2: Recommended electrical symbols. (cont'd)

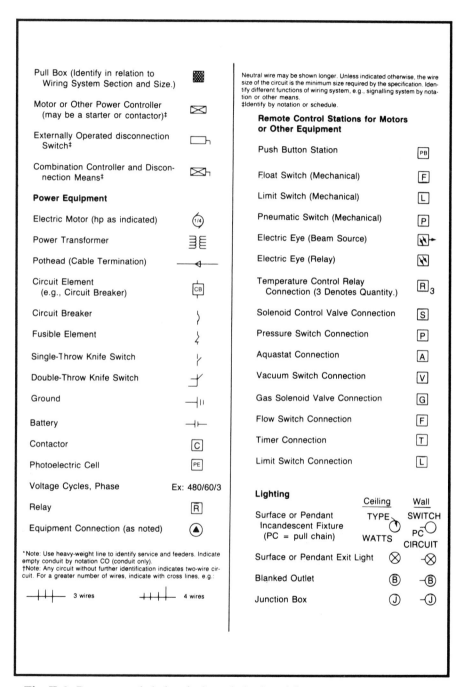

Pull Box (Identify in relation to Wiring System Section and Size.)

Motor or Other Power Controller (may be a starter or contactor)‡

Externally Operated disconnection Switch‡

Combination Controller and Disconnection Means‡

Power Equipment

Electric Motor (hp as indicated)

Power Transformer

Pothead (Cable Termination)

Circuit Element (e.g., Circuit Breaker)

Circuit Breaker

Fusible Element

Single-Throw Knife Switch

Double-Throw Knife Switch

Ground

Battery

Contactor

Photoelectric Cell

Voltage Cycles, Phase — Ex: 480/60/3

Relay

Equipment Connection (as noted)

*Note: Use heavy-weight line to identify service and feeders. Indicate empty conduit by notation CO (conduit only).
†Note: Any circuit without further identification indicates two-wire circuit. For a greater number of wires, indicate with cross lines, e.g.:

3 wires   4 wires

Neutral wire may be shown longer. Unless indicated otherwise, the wire size of the circuit is the minimum size required by the specification. Identify different functions of wiring system, e.g., signalling system by notation or other means.
‡Identify by notation or schedule.

Remote Control Stations for Motors or Other Equipment

Push Button Station

Float Switch (Mechanical)

Limit Switch (Mechanical)

Pneumatic Switch (Mechanical)

Electric Eye (Beam Source)

Electric Eye (Relay)

Temperature Control Relay Connection (3 Denotes Quantity.)

Solenoid Control Valve Connection

Pressure Switch Connection

Aquastat Connection

Vacuum Switch Connection

Gas Solenoid Valve Connection

Flow Switch Connection

Timer Connection

Limit Switch Connection

Lighting

Surface or Pendant Incandescent Fixture (PC = pull chain)

Surface or Pendant Exit Light

Blanked Outlet

Junction Box

Fig. E-2: Recommended electrical symbols. (cont'd)

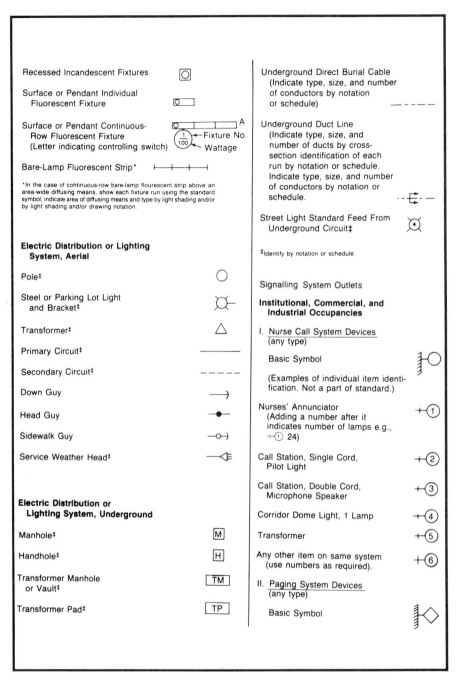

Recessed Incandescent Fixtures

Surface or Pendant Individual
Fluorescent Fixture

Surface or Pendant Continuous-
Row Fluorescent Fixture
(Letter indicating controlling switch)
Fixture No.
Wattage

Bare-Lamp Fluorescent Strip*

*In the case of continuous-row bare-lamp flourescent strip above an
area-wide diffusing means, show each fixture run using the standard
symbol; indicate area of diffusing means and type by light shading and/or
by light shading and/or drawing notation.

**Electric Distribution or Lighting
System, Aerial**

Pole‡

Steel or Parking Lot Light
and Bracket‡

Transformer‡

Primary Circuit‡

Secondary Circuit‡

Down Guy

Head Guy

Sidewalk Guy

Service Weather Head‡

**Electric Distribution or
Lighting System, Underground**

Manhole‡

Handhole‡

Transformer Manhole
or Vault‡

Transformer Pad‡

Underground Direct Burial Cable
(Indicate type, size, and number
of conductors by notation
or schedule)

Underground Duct Line
(Indicate type, size, and
number of ducts by cross-
section identification of each
run by notation or schedule.
Indicate type, size, and number
of conductors by notation or
schedule.

Street Light Standard Feed From
Underground Circuit‡

‡Identify by notation or schedule.

Signalling System Outlets

**Institutional, Commercial, and
Industrial Occupancies**

I. Nurse Call System Devices
(any type)

Basic Symbol

(Examples of individual item identification. Not a part of standard.)

Nurses' Annunciator
(Adding a number after it
indicates number of lamps e.g.,
+① 24)

Call Station, Single Cord,
Pilot Light

Call Station, Double Cord,
Microphone Speaker

Corridor Dome Light, 1 Lamp

Transformer

Any other item on same system
(use numbers as required).

II. Paging System Devices
(any type)

Basic Symbol

Fig. E-2: Recommended electrical symbols. (cont'd)

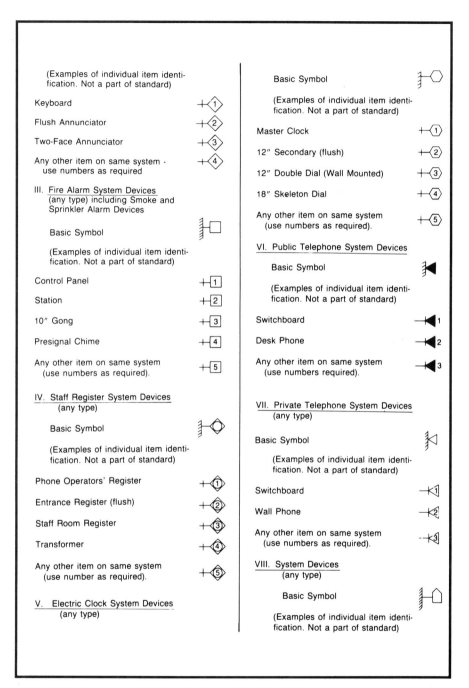

(Examples of individual item identification. Not a part of standard)

Keyboard

Flush Annunciator

Two-Face Annunciator

Any other item on same system -
use numbers as required

III. Fire Alarm System Devices
(any type) including Smoke and
Sprinkler Alarm Devices

Basic Symbol

(Examples of individual item identification. Not a part of standard)

Control Panel

Station

10″ Gong

Presignal Chime

Any other item on same system
(use numbers as required).

IV. Staff Register System Devices
(any type)

Basic Symbol

(Examples of individual item identification. Not a part of standard)

Phone Operators' Register

Entrance Register (flush)

Staff Room Register

Transformer

Any other item on same system
(use number as required).

V. Electric Clock System Devices
(any type)

Basic Symbol

(Examples of individual item identification. Not a part of standard)

Master Clock

12″ Secondary (flush)

12″ Double Dial (Wall Mounted)

18″ Skeleton Dial

Any other item on same system
(use numbers as required).

VI. Public Telephone System Devices

Basic Symbol

(Examples of individual item identification. Not a part of standard)

Switchboard

Desk Phone

Any other item on same system
(use numbers required).

VII. Private Telephone System Devices
(any type)

Basic Symbol

(Examples of individual item identification. Not a part of standard)

Switchboard

Wall Phone

Any other item on same system
(use numbers as required).

VIII. System Devices
(any type)

Basic Symbol

(Examples of individual item identification. Not a part of standard)

Fig. E-2: Recommended electrical symbols. (cont'd)

electricity: Relating to the flow or presence of charged particles; a fundamental physical force or energy.

electric sign: A fixed,, stationary, or portable self-contained, electrically illuminated utilization equipment with words or symbols designed to convey information or attract attention.

electric water valve: Solenoid type (electrically operated) valve used to turn water flow on and off.

electrocution: Death caused by electrical current through the heart, usually in excess of 50 ma.

electrode: A conductor through which current transfers to another material.

electrolysis: The production of chemical changes by the passage of current from an electrode to an electrolyte or vice versa.

electrolyte: A liquid or solid that conducts electricity by the flow of ions.

electrolytic condenser-capacitor: Plate or surface capable of storing small electrical charges. Common electrolytic condensers are formed by rolling thin sheets of foil between insulating materials. Condenser capacity is expressed in microfarads.

electromagnet: A device consisting of a ferromagnetic core and a coil that produces appreciable magnetic effects only when an electric current exists in the coil.

electromotive force (emf) voltage: Electrical force that causes current (free electrons) to flow or move in an electrical circuit. Unit of measurement is the volt.

electron emission: The release of electrons from the surface of a material into surrounding space due to heat, light, high voltage, or other causes.

electron: The subatomic particle that carries the unit negative charge of electricity.

electronegative gas: A type of insulating gas used in pressure cables, such as SF6.

electronics: The science dealing with the development and application of devices and systems involving the flow of electrons in vacuum, gaseous media, and semi-conductors.

electro-osmosis: The movement of fluids through diaphragms because of electric current.

electroplating: Depositing a metal in an adherent form upon an object using electrolysis.

electropneumatic: An electrically controlled pneumatic device.

electrostatics: Electrical charges at rest in the frame of reference.

electrotherapy: The use of electricity in treatment of disease.

electrothermics: Direct transformations of electric and heat energy.

electrotinning: Depositing tin on an object.

elevation: Drawing showing the projection of a building on a vertical plane.

elongation: 1) The fractional increase in length of a material stressed in tension. 2) The amount of stretch of a material in a given length before breaking.

EMA (Electrical Moisture Absorption): A water tank test during which the sample cables are subjected to voltage while the water is maintained at rated temperature; the immersion time is long, in order to accelerate failure due to moisture in the insulation; simulates buried cable.

EMI: Electromagnetic interference.

emitter: The part of a transistor that emits electrons.

emulsifying agent: A material that increases the stability of an emulsion.

emulsion: The colloidal suspension of one liquid in another liquid, such as oil in water for lubrication.

enameled wire: Wire insulated with a thin baked-on varnish enamel, used in coils to allow the maximum number of turns in a given space.

enclosed: Surrounded by a case, housing, fence, or wall which will prevent persons from accidentally contacting energized parts.

enclosure: The case or housing of apparatus, or the fence or walls surrounding an installation to prevent personnel from accidentally contacting energized parts, or to protect the equipment from physical damage.

energy: The ability to do work; such as heat, light, electrical, mechanical, etc.

engine: An apparatus that converts heat to mechanical energy.

environment: 1) The universe within which a system must operate. 2) All the elements over which the designer has no control and that affect a system or its inputs.

EPA (Environmental Protection Agency): The federal regulatory agency responsible for maintaining and improving the quality of our living environment — mainly air and water.

epitaxial: A very significant thin-film type of deposit for making certain devices in microcircuits involving a realignment of molecules.

EPRI (Electric Power Research Institute): An organization to develop and manage a technology for improving electric power production, distribution and utilization; sponsored by electric utilities.

equilibrium: Properties are time constant.

equipment: A general term including material, fittings, devices, appliances, fixtures, apparatus, and the like used as part of, or in connection with, an electrical installation.

equipment grounding conductor: The conductor used to connect the noncurrent-carrying metal parts of equipment, raceways, and other enclosures to the system grounded conductor and/or the grounding electrode conductor at the service equipment or at the source of a separately derived system.

equipotential: Having the same voltage at all points.

equivalent circuit: An arrangement of circuit elements that has characteristics over a range of interest electrically equivalent to those of a different circuit or device.

ERDA (Energy Research & Development Administration): Federal agency (replacing part of AEC) for research and relating to energy — new sources, better efficiency, etc.

erosion: Destruction by abrasive action of fluids.

escutcheon: The plate about a keyhole or the one to which a door knocker is attached.

estimating: Calculating the amount of material required for a piece of work; also the labor required and determining the value of the finished product.

etching: Revealing structural details of a metal surface using chemical or electrolytic action.

ETL: Electrical Testing Laboratory.

evaporation: A term applied to the changing of a liquid to a gas; heat is absorbed in this process.

evaporator: Part of a refrigerating mechanism in which the refrigerant vaporizes and absorbs heat.

excavation: A digging out of earth to make room for engineering improvements. A cavity is so formed.

excitation losses: Losses in a transformer or electrical machine because of voltage.

excite: To initiate or develop a magnetic field.

expansion bolt: Bolt with a casing arranged to wedge the bolt into a masonry wall to provide an anchor.

expansion joint: Joint between two adjoining concrete members arranged to permit expansion and contraction with temperature changes.

expansion, thermal: The fractional change in unit length per unit temperature change.

expansion valve: A device in a refrigerating system that maintains a pressure difference between the high side and low side and is operated by pressure.

explosionproof: Designed and constructed to withstand an internal explosion without creating an external explosion or fire. Several explosionproof fittings are shown in Fig E-3.

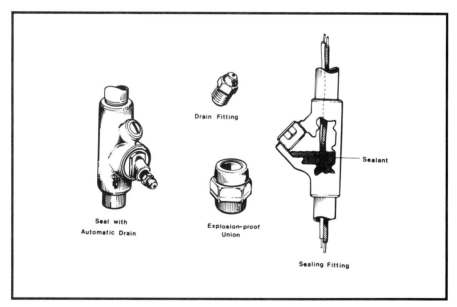

Fig. 3: Several types of explosionproof fittings.

explosionproof apparatus: Apparatus enclosed in a case that is capable of withstanding an explosion of a specified gas or vapor that may occur within it, and also capable of preventing the ignition of a specified gas or vapor surrounding the enclosure by sparks, flashes, or explosion of the gas or vapor within, and which operates at such an external temperature that a surrounding flammable atmosphere will not be ignited thereby.

exponential: Pertaining to the mathematical expression, $y = aebx$.

exposed (as applied to live parts): Live parts that a person could inadvertently touch or approach nearer than a safe distance. This term is applied to parts not suitably guarded, isolated, or insulated.

exposed (as applied to wiring method): Not concealed; externally operable; capable of being operated without exposing the operator to contact with live parts.

extender: A substance added to a plastic to reduce the amount of the primary resin required per unit area.

exterior: The outside of, as a whole or in part, as the exterior of a building; an exterior wall.

externally operable: Capable of being operated without exposing the operator to contact with live parts.

extraction: The transfer of a material from a substance to a liquid in contact with the substance.

extrude: To form materials to a given cross section by forcing through a die.

extruder types: 1) Strip — uses strips of compound. 2) Powder/pellet — uses compound in powder or pellet form.

eyelet: Something used on printed circuit boards to make reliable connections from one side of the board to the other.

F

fabrication: The act of building or putting together. Forming into a whole by uniting or assembling of parts.

facade: Main front of a building.

face: An operation that machines the sides or ends of the piece.

face brick: Brick selected for appearance in an exposed wall.

face of a gear: That portion of the tooth curve above the pitch circle and measured across the rim of the gear from one end of the tooth to the other.

facsimile: The remote reproduction of graphic material; an exact copy.

factorial experiment: Having more than one factor as a controlled variable in one experiment; produces much data per experiment, but the results are complex to analyze.

factor of safety: Ratio of ultimate strength of material to maximum permissible stress in use.

Fahrenheit: The commonly used thermometer scale named after Gabriel Fahrenheit, a German physicist (1686-1736); the freezing point is 32°, and the boiling point 212°.

fail-safe control: A device that opens a circuit when the sensing element fails to operate.

failure: Termination of the ability of an item to perform the required function.

fan: A radial or axial flow device used for moving or producing artificial currents of air.

FAO: This symbol on a mechanical drawing means that the piece is machined or finished all over.

farad: The basic unit of capacitance; 1 farad equals one coulomb per volt: $1f = $ m-2.kg-1s4.A2. It is too large a unit for practical work. Its common use is in terms of "microfarad."

Faraday, Michael: (1791-1867). English scientist; early investigator and experimenter who invented the transformer.

Faraday's laws of electrolysis: 1) The weight of the products of electrolysis is proportional to the quantity of electricity that has passed through the electrolyte. 2) For a given quantity of electricity the weight of the products of electrolysis is proportional to their electrochemical equivalents.

fatigue: The weakening or breakdown of a material due to cyclic stress.

fatigue strength: The maximum stress that can be sustained for a specified number of cycles without failure, the stress being completely reversed within each cycle unless otherwise stated.

fault: An abnormal connection in a circuit.

fault, arcing: A fault having high impedance causing arcing.

fault, bolting A fault of very low impedance.

fault, ground: a fault to ground.

FCC: Flat Conductor Cable. Type FCC consists of three or more flat copper conductors placed edge-to-edge and separated and enclosed within an insulating assembly.

feedback: The process of transferring energy from the output circuit of a device back to its input.

feeder: A circuit, such as conductors in conduit or a busway run, which carriers a large block of power from the service equipment to a sub-feeder panel or a branch circuit panel or to some point at which the block or power is broken down into smaller circuits.

Ferranti effect: When the voltage is greater than the source voltage in an ac cable or transmission line.

fiber optics: Transmission of energy by light through glass fibers.

fibrillation: A continued, uncoordinated activity in the fibers of the heart, diaphragm, or other muscles.

fiddle: A small, hand-operated drill.

fidelity: The degree to which a system accurately reproduces an input.

field: The effect produced in surrounding space by an electrically charged object, by electrons in motion, or by a magnet.

field coil: The coil or winding around the field magnets or pole pieces of a motor or generator.

field core: The iron projection usually salient, upon which is wound the field winding of a generator or motor.

field density: The density of the magnetic field or the magnetic flux, measured in the number of lines of force per unit area, is dependent upon the strength of the field element, the number of turns of wire, and the size and characteristics of the pole piece.

field distortion: The distortion of the normal field existing between the north and south poles of a generator due to the counter electromotive force generated in the armature windings.

field, electrostatic: The region near a charged object.

field excitation: The magnetic effect produced in an electromagnet when current is passed through a winding, usually with an iron core.

field magnet: The electromagnet by which the magnetic field of force is produced in a generator.

field rheostat: A variable high resistance of low-current capacity inserted in the field circuit to regulate within limits the output of a generator.

filament: A cathode in the form of a metal wire in an electron tube; a thin wire or fiber.

filled tape: A fabric tape having interstices, but not necessarily the surface, filled with a compound to prevent wicking, improve strength, make conductive, etc.

filler: A cheap and relatively inert substance added to plastic or rubber to make it less costly and improve physical properties.

filler, cable: Materials used to fill voids and spaces in a cable construction; normally to give a smooth outer configuration, and also may serve as flame retardants, etc.

fillet: The rounded corner or portion that joins two surfaces at an angle to each other.

film: A rectangular product having thickness of 0.010 inch thick or less.

filter: A porous article through which a gas or liquid is passed to separate out matter in suspension; a circuit or devices that pass one frequency or frequency band while blocking others, or vice versa.

final: The final inspection of an electrical installation, e.g., "The contractor got the final on the job."

final tests: Those performed on the completed cable (after manufacturing).

fines: Fill material such as rocks having ⅛ inch as the largest dimension.

finish plaster: Final or white coat of plaster.

firebrick: Brick made to withstand high temperatures and used for lining chimneys, incinerators, and similar structures.

fireproof wood: Chemically treated wood; fire-resistive, used where incombustible materials are required.

fire-rated doors: Doors designed to resist standard fire tests and labeled for identification.

fire-resistance rating: The time in hours the material or construction will withstand fire exposure as determined by certain standards.

fire-shield cable: Material or devices to prevent fire spread between raceways.

fire-stop: A barrier to minimize fire spread.

fish: To fish wire or cable means to pull it through conduit, raceway or other confined spaces, like walls or ceilings.

fish tape: A flexible metal tape for fishing through conduits or other raceway to pull in wires or cables; also made in non-metallic form of "rigid rope" for hand fishing of raceways. See Fig. F-1.

fission: (nuclear power): The splitting of an atom into two fragments by bombarding its nucleus with particles releasing high kinetic energy (32pj) and two or three neutrons along with radiation; the most important type of fission is that caused by neutrons because it can be self-sustaining due to chain reactions — the newly released neutrons can cause other fissions to occur.

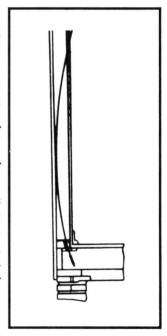

Fig. F-1: Fish tape used to pull wires through partition.

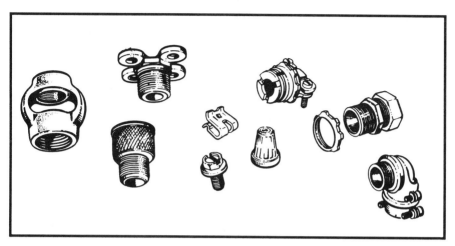

Fig. F-2: Several types of fittings that are used in a mechanical way rather than electrically on electrical systems.

fitting: An accessory such as a locknut, bushing, or other part of a wiring system that is intended primarily to perform a mechanical rather than an electrical function. See Fig. F-2.

five hundred thousands: Referring to size of conductors by their MCM or current kcmil rating, e.g., "Two hundred and fifty thousands" is a number of 250 kcmil conductors; "Twin three hundred thousands" is two conductors of 300,000 circular mil size. The 1990 NE Code now refers to kcmil instead of the former MCM.

fixture splice: The fixture wire is wound in close turns around the main conductor after which the end of the main conductor is bent tightly back over the coils of fixture wire.

fixture wire: Usually 16 or 18 gauge, solid or stranded and insulated. It is used for wiring electric fixtures.

flag: A visual indicator for event happenings such as the activation and reclosing of an automatic circuit breaker.

flame-retardant: 1) Does not support or convey flame. 2) An additive for rubber or plastic that enhances its flame resistance.

flange: The circular disks on a reel to support the drum and keep the cable on the drum.

flapper valve: The type of valve used in refrigeration compressors that allow gaseous refrigerants to flow in only one direction.

flasher: A device for periodically flashing on and off a lamp or group of lamps, as in some electric signs.

flashover: A momentary electrical interconnection around or over the surface of an insulator.

flashpoint: The lowest temperature at which a combustible substance ignites in air when exposed to flame.

flat: Of uniform thickness; eliminates the drops of beams and girders.

flat wire: A rectangular wire having 0.188 inch thickness or less, 1¼-inch width or less.

Fleming's rule: Right-hand rule. If the forefinger points along the lines of flux, and the thumb in the direction of the motion of the conductor, the middle finger will point in the direction of the induced e.m.f. Left-hand rule: Point the forefinger in the direction of the flux, the middle finger in the direction of the current in the conductor, then the thumb will point in the direction in which the conductor tends to move. Fleming's rules determine the direction in which a motor will rotate, or the polarity of a generated current from a generator.

Flemish bond: Pattern of bonding in brickwork consisting of alternate headers and stretchers in the same course.

flex: Common term used to refer to flexible metallic conduit. See *Greenfield.*

flexural strength: The strength of a material in bending, expressed as the tensile stress of the outermost fibers of a bent test sample at the instant of failure.

flitch beam: Built-up beam consisting of a steel plate sandwiched between wood members and bolted.

floating: Not having a distinct reference level with respect to voltage measurements. For example, a "floating neutral" on a balanced three-wire delta system.

float switch: A switch that is opened and closed by a float that rises and falls with the level of the liquid in a tank.

float valve: Type of valve that is operated by a sphere or pan that floats on a liquid surface and controls the level of liquid.

flooding: Act of filling a space with a liquid.

flow meter: Instrument used to measure velocity or volume of fluid movement.

fluctuation: A variation, an irregular change of movement.

flux: 1) The rate of flow of energy across or through a surface. 2) A substance used to promote or facilitate soldering or welding by removing surface oxides.

foamed insulation: Insulation made sponge-like by using foaming or blowing agents to create the cells.

foil: Metal film.

footing: Structural unit used to distribute loads to the bearing materials.

forced convection: Movement of fluid by mechanical force such as fans or pumps.

foundation: Composed of footings, piers, foundation walls (basement walls), and any special underground construction necessary to properly support the structure.

FPM: Feet per minute.

FR-1: See VW-1.

frequency: The number of complete cycles an alternating electric current, sound wave, or vibrating object undergoes per second.

friction pile: Pile with supporting capacity produced by friction with the soil in contact with the pile.

friction tape: An insulating tape made of asphalt impregnated cloth; used on 600V cables.

frost line: Deepest level below grade to which frost penetrates in a geographic area.

fuel cell: A cell that can continually change chemical energy to electrical energy.

full braid: One made of a single material as opposed to one of a mixture of materials.

function: A quantity whose value depends upon the value of another quantity.

furring: Thin wood, brick, or metal applied to joists, studs, or wall to form a level surface (as for attaching wallboard) or airspace.

fuse: A protecting device that opens a circuit when the fusible element is severed by heating, due to overcurrent passing through. Rating: voltage, normal current, maximum let-thru current, time delay of interruption.

fuse clips: The spring part of a cutout or switch that holds the ferrules of a cartridge fuse.

fuse, dual element: A fuse having two fuse characteristics; the usual combination is having an overcurrent limit and a time delay before activation. See Fig. F-3.

fuse link: The fusible part of a cartridge fuse.

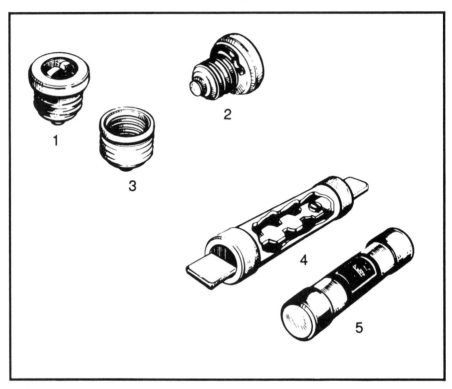

Fig. F-3: Fuses are manufactured in a variety of shapes and sizes: 1) Edison-base plug fuse; 2) Type S plug fuse; 3) Type S plug-fuse adapter; 4) knife-type cartridge fuse with renewable element; 5) ferrule-type cartridge fuse with non-renewable element.

fuse, nonrenewable or one-time: A fuse that must be replaced after it interrupts a circuit.

fuse, renewable link: A fuse that may be reused after current interruption by replacing the meltable link.

fuse wire: Wire made of an alloy that melts at a low temperature.

fusible plug: A plug or fitting made with a metal of a known low melting temperature; used as a safety device to release pressures in case of fire.

G

gable: The triangular end of an exterior wall above the eaves.

gable roof: A ridge roof terminating in a gable. See Fig. G-1.

gain: 1) The ratio of output to input power, voltage, or current, respectively. 2) The increase in signal power produced by an amplifier.

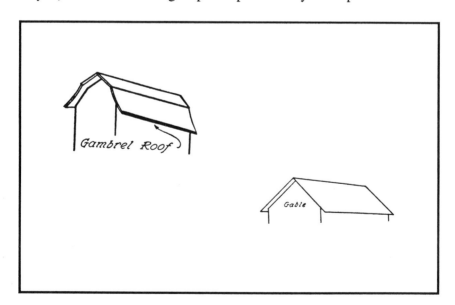

Fig. G-1: Roof styles.

galvanic action: The action upon one another of electropositive and electronegative metals, wasting away the positive metal; particularly noticeable where iron and copper, or zinc and copper, or brass and iron, are in contact in the presence of acidulated water.

galvanometer: An instrument for indicating or measuring a small electrical current by means of a mechanical motion derived from electromagnetic or dynamic forces.

gambrel roof: Roof with its slope broken by an obtuse angle. See Fig. G-1 on page 73.

gang switch: A unit of two or more switches to give control of two or more circuits from one point. The entire mechanism is mounted in one box under one cover.

garage: A building or portion of a building in which one or more self-propelled vehicles carrying volatile, flammable liquid for fuel or power are kept.

garden bond: Bond formed by inserting headers at wide intervals.

gas: Vapor phase or state of a substance.

gas filled pipe cable: See *pressure cable.*

gas pocket: A cavity caused by entrapped gas.

gate: A device with multiple inputs and a single output that makes an electronic circuit operable for a short time. There are five basic types of gates: *and, or, nand, nor,* and *inverter.*

gauge: 1) Dimension expressed in terms of a system of arbitrary reference numbers; dimensions expressed in decimals are preferred. 2) To measure.

gem box: The most common rectangular outlet box used to hold wall switches and receptacle outlets installed recessed in walls; made in wide variety on constructions: 2 in. wide by 3 in. high by various depths; without clamps for conduit and with or without clamps for cable (armored or nonmetallic sheathed); single-gang boxes that can be ganged together for more than one device.

generator: 1) A rotating machine that is used to convert mechanical to electrical energy. 2) Automotive-mechanical to direct current. 3) General-apparatus, equipment, etc. to convert or change energy from one form to another.

geometric factor: A parameter used and determined solely by the relative dimensions and configuration of the conductors and insulation of a cable.

GFCI (Ground-Fault Circuit-Interrupter): A protective device that detects abnormal current flowing to ground and then interrupts the circuit.

GFPE (Ground-Fault Protection of Equipment): A system intended to provide protection of equipment from damaging line-to-ground fault currents by operating to cause a disconnecting means to open all ungrounded conductors of the faulted circuit.

girder: A large beam made of wood, steel, or reinforced concrete.

girt: Heavy timber framed into corner posts as support for the building.

government anchor: A V-shaped anchor usually made of ½-inch round bars to secure the steel beam to masonry.

grade beam: Horizontal, reinforced concrete beam between two supporting piers at or below ground supporting a wall or structure.

graded insulation: Combining insulations in a manner to improve the electric field distribution across the combination.

gradient: The rate of change of a variable magnitude.

grain: An individual crystal in a polycrystalline metal or alloy.

gray: The derived SI unit for absorbed radiation dose; one gray equals one joule per kilogram.

Greenfield: Another name used to refer to flexible metal conduit. See Fig. G-2.

grid: An electrode having one or more openings for the passage of electrons or ions.

grid leak: A resistor of high ohmic value connected between the control grid and the cathode in a grid-leak capacitor detector circuit and used for automatic biasing.

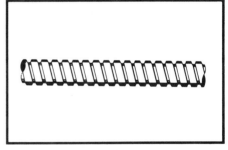

grillage: Steel framework in a foundation designed to spread a concentrated load over a wider area; generally enclosed in concrete.

Fig. G-2: Greenfield or flexible metallic conduit.

grille: An ornamental or louvered opening placed at the end of an air passageway.

groined ceiling: Arched ceiling consisting of two intersecting curves, planes.

grommet: A plastic, metal, or rubber doughnut-shaped protector for wires or tubing as they pass through a hole in an object.

ground: A large conducting body (as the earth) used as a common return for an electric circuit and as an arbitrary zero of potential.

ground check: A pilot wire in portable cables to monitor the grounding circuit.

ground clamp: A clamp used for attaching a wire or other conductor to a pipe to make a good electrical connection.

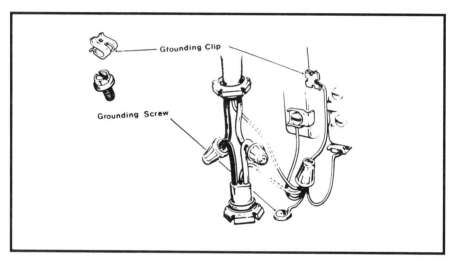

Fig. G-3: A ground clip and its practical application.

ground clip: A spring clip used to secure a bonding conductor to an outlet box. See Fig. G-3.

ground coil: A heat exchanger buried in the ground that may be used either as an evaporator or a condenser.

ground conductor: A conductor used to connect equipment or the grounded circuit of a wiring system to a grounding electrode or electrodes.

grounded: Connected to earth or to some conducting body that serves in place of the earth.

grounded conductor: A system or circuit conductor that is intentionally grounded, usually gray or white in color.

ground-fault protection of equipment: See *GFPE.*

grounding: The device or conductor connected to ground designed to conduct only in abnormal conditions.

grounding conductor: A conductor used to connect metal equipment enclosures and/or the system grounded conductor to a grounding electrode, such as the ground wire run to the water pipe at a service; also may be a bare or insulated conductor used to ground motor frames, panel boxes, and other metal equipment enclosures used throughout an electrical system. In most conduit systems, the conduit is used as the ground conductor. See Fig. G-4.

grounds: Narrow strips of wood nailed to walls as guides to plastering and as a nailing base for interior trim.

group ambient temperature: The no-load temperature of a cable group with all other cables or ducts loaded.

guard: 1) A conductor situated to conduct interference to its source and prevent its influence upon the desired signal. 2) Covered, shielded, fenced, enclosed, or otherwise protected by means of suitable covers, casings, barriers, rails, screens, mats, or platforms to remove the likelihood of approach or contact by persons or objects to a point of danger.

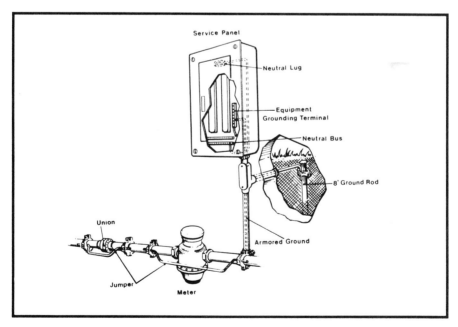

Fig. G-4: One NE Code approved method of grounding an electric service.

guider: The adjustable part of the mold of an extruder.

gusset: A plate or bracket for strengthening an angle in framework.

gutter: The space provided along the sides and at the top and bottom of enclosures for switches, panels, and other apparatus, to provide for arranging conductors that terminate at the lugs or terminals of the enclosed equipment. Gutter is also used to refer to a rectangular sheet metal enclosure with removable side, used for splicing and tapping wires at distribution centers and motor control layouts. Usually called "auxiliary gutter."

guy: A tension wire connected to a tall structure such as a power pole and another fixed object (anchor) to add strength to the structure.

H

hack saw: A light-framed saw used for cutting metal conduit, BX cable, and other metal objects.

half hard: A relative measure of conductor temper.

half lap, joint: Joint formed by cutting away half the thickness of each piece.

half wave: Rectifying only half of a sinusoidala ac supply. See Fig. H-1.

Hall effect: The changing of current density in a conductor due to a magnetic field extraneous to the conductor.

halogen: "Salt former." Applied to the family of elements consisting of fluorine, chlorine, bromine, and iodine.

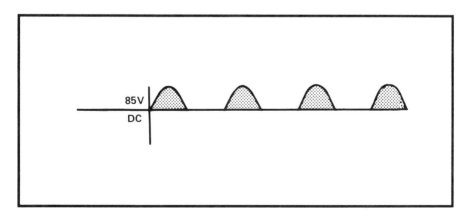

Fig. H-1: Half-wave depicted graphically.

handhole: A small box in a race-way used to facilitate cable installation into which workers reach but do not enter. See Fig. H-2.

handy box: The commonly used, single-gang outlet box used for surface mounting to enclose receptacles or wall switches on concrete or cinder block construction of industrial and commercial buildings; non-gangable; also made for recessed mounting; also known as "utility boxes". See Fig. H-3 on page 81.

hard drawn: A relative measure of temper; drawn to obtain maximum strength.

hardness: Resistance to plastic deformation usually by deformation; stiffness or temper; resistance to scratching, abrasion or cutting.

harmonic: An oscillation whose frequency is an integral multiple of the fundamental frequency.

harness: A group of conductors laced or bundled in a given configuration, usually with many breakouts.

hat: A special pallet for transporting long rubber strips or coils of wire; the pallets look like a hat.

hazardous (classified) location: A location in which ignitable vapors, dust, or fibers may cause fire or explosion.

H beam: Steel beam with wider flanges than an I beam.

HDP: High density polyethylene.

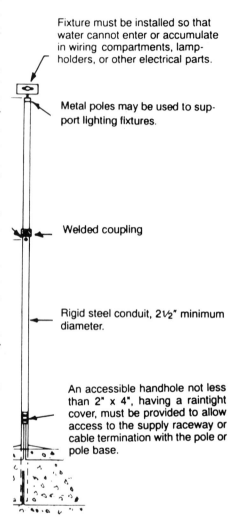

Fixture must be installed so that water cannot enter or accumulate in wiring compartments, lamp-holders, or other electrical parts.

Metal poles may be used to support lighting fixtures.

Welded coupling

Rigid steel conduit, 2½" minimum diameter.

An accessible handhole not less than 2" x 4", having a raintight cover, must be provided to allow access to the supply raceway or cable termination with the pole or pole base.

Fig. H-2: Handhole in lighting standard to facilitate wire connections.

header: 1) Brick laid with an end exposed in the wall. 2) wood beam set between two trimmers and carrying the tail beams. 3) A transverse raceway for electric conductors, providing access to predetermined cells of a cellular metal floor, thereby permitting the installation of electric conductors from a distribution center to the cells.

heat: A fundamental physical force or energy relating to temperature.

heat dissipation: The flow of heat from a hot body to a cooler body by convection, radiation, or conduction.

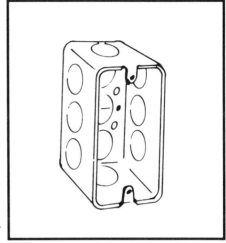

Fig. H-3: Utility or "handy" box.

heat exchanger: A device used to transfer heat from a warm or hot surface to a cold or cooler surface. Evaporators and condensers are heat exchangers.

heating unit: The part of any electrical heating device in which the heat is generated.

heating valve: Amount of heat that may be obtained by burning a fuel; usually expressed in Btu per pound or gallon.

heat load: Amount of heat, measured in Btu, that is removed during a period of 24 hours.

heat pump: A compression cycle system used to supply heat to a temperature—controlled space, which can also remove heat from the same space.

heat sink: A part used to absorb heat from another device.

heat transfer: Movement of heat from one body or substance to another. Heat may be transferred by radiation, conduction, convection, or a combination of these.

heat treatment: Heating and cooling a solid metal or alloy to obtain desired properties or conditions; excluding heating for hot work.

heavy water (nuclear power): Heavy water, D^2O, contains deuterium which is hydrogen atoms having twice the ordinary mass.

helix: The path followed when winding a wire or strip around a tube at a constant angle.

henry: The derived SI unit for inductance; one henry equals one weber per ampere. A circuit has an inductance of one henry when a current changing at the rate of one ampere per second induces one volt in the circuit.

hermetic motor: A motor designed to operate within refrigeration fluid. See NE Code Section 440-2.

hertz: The derived SI unit for frequency; one hertz equals one cycle per second.

hickey: 1) A conduit bending tool. 2) A box fitting for hanging lighting fixtures. See Fig. H-4.

high-hat: A ceiling recessed incandescent lighting fixture of round cross-section, looking like a man's high hat in the shape of its construction. See Fig. H-5 on page 83.

high-leg: That phase conductor of a 3-phase, 4-wire, delta-connected system that is not connected to the single-phase power supply; the conductor with the highest voltage to ground; this phase conductor must be identified (per NE Code) and is commonly painted orange to provide such identification.

high pressure cutout: Electrical control switch operated by the high side pressure that automatically opens an electrical circuit if too high head pressure or condensing pressure is reached.

high side: Parts of a refrigerating system that are under condensing or high side pressure.

hi-pot test: A high-potential test in which equipment insulation is subjected to voltage level higher than that for which it is rated to find any weak spots or deficiencies in the insulation.

HMP, HMPE: High molecular weight polyethylene.

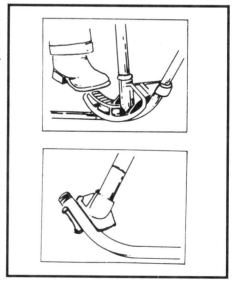

Fig. H-4: Pipe-bending tool for rigid steel conduit; normally called "hickey."

hole: A mobile vacancy in the electron structure of semi-conductors that acts like a positive electron charge with mass.

home run: That part of a branch circuit from the panelboard housing the branch circuit fuse or CB (overcurrent protection) and the first junction box at which the branch circuit is spliced to lighting or receptacle devices or to conductors that continue the branch circuit to the next outlet or junction box. The term "home run" is usually reserved for multi-outlet lighting and appliance circuits. See Fig. H-6 on page 84.

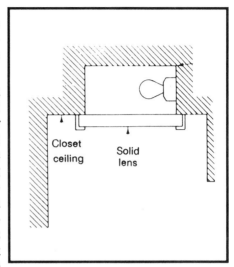

Fig. H-5: Recessed lighting fixture, sometimes called "high-hat."

horsepower: The non-SI unit for power; 1 hp = 1 HP = 746 w (electric) = 9800 w (boiler).

hot: Energized with electricity.

hot dip: Coating by dipping into a molten bath.

hot gas bypass: Piping system in a refrigerating unit that moves hot refrigerant gas from a condenser into the low pressure side.

hot junction: That part of the thermoelectric circuit that releases heat.

hot leg: A circuit conductor that normally operates at a voltage above ground; the phase wires or energized circuit wires other than a grounded neutral wire or grounded phase leg. The ungrounded conductor of a circuit.

hot modulus: Stress at 100% elongation after 5 minutes of conditioning at a given temperature (normally 130°C).

hot stick: A long insulated stick having a hook at one end that is used to open energized switches, etc.

hot wire: A resistance wire in an electrical relay that expands when heated and contracts when cooled.

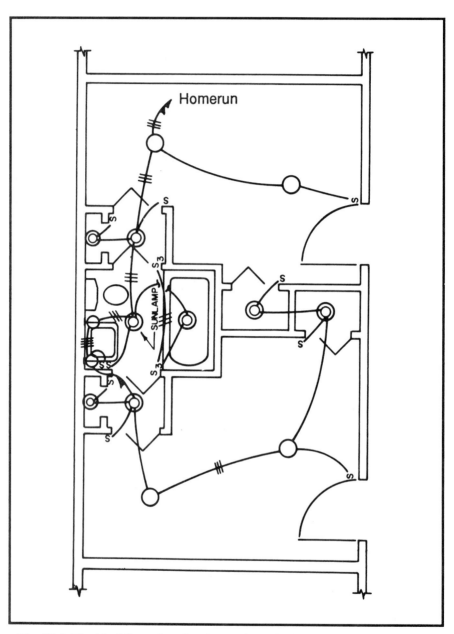

Fig. H-6: Electrical floor plan showing outlets, circuits, and homeruns; the latter is indicated by half-arrowheads.

HTGR (Hi-Temp, Gas-Cooled Reactor) (nuclear power): A basic nuclear power fission reactor in which the reactor heats a gas; the gas exchanges its heat with a secondary loop to produce steam for the turbine or generator.

hub: 1) A fitting to attach threaded conduit to boxes. See Fig. H-7. 2) The central part of a cylinder into which a shaft may be inserted. 3) A reference point used for overhead line layout.

hum: Interference from ac power, normally of low frequency and audible.

humidity: Moisture, dampness. Relative humidity is the ratio of the quantity of vapor present in the air to the greatest amount possible at a given temperature.

hydraulic jack: A lifting jack, actuated by a small force pump enclosed within it, and operated by a lever from the outside.

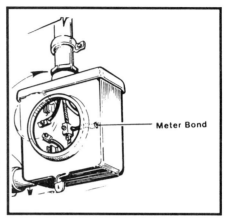

Fig. H-7: Electric meter base with hub for conduit or WP SE cable connectors.

hydrometer: Floating instrument used to measure specific gravity of a liquid. Specific gravity is the ratio of the density of a material to the density of a substance accepted as a standard. See Fig. H-8.

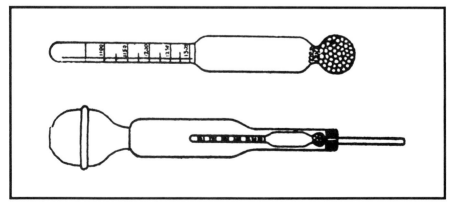

Fig. H-8: Hydrometer is used to measure specific gravity of a liquid, especially that in a storage battery.

hydronic: Type of heating system that circulates a heated fluid, usually water, through baseboard coils. Circulation pump is usually controlled by a thermostat.

hygrometer: An instrument used to measure the degree of moisture in the atmosphere.

hygroscopic: Readily absorbing and retaining moisture.

Hypalon® The Dupont trade name for chlorosulfonated polyethylene.

Hypot® Registered trade name by Associated Research, Inc. for their hi-pot tester.

hysteresis: The time lag exhibited by a body in reacting to changes in forces affecting it; an internal friction.

I

I^2r: The power losses in a device or apparatus due to resistance heating.

I^2t: Relating to the heating effect of a current (amps-squared) for a specified time (seconds), under specified conditions.

IACS: (International Annealed Copper Standard): Refined copper for electrical conductors; 100% conductivity at 20C for 1m x 1mm2 has ⅟₅₈ ohm resistivity, 8.89 grams per mm2 density, 0.000 017 per degree C coefficient of linear expansion, ⅟₂₅₄.₄₅ per degree C coefficient of variation of resistance; however, NBS suggest the use of 8.93 for density (1977).

IAEI: International Association of Electrical Inspectors.

I beam: Rolled steel beam or built-up beam of an I section. See Fig. I-1.

IBEW: International Brotherhood of Electrical Workers.

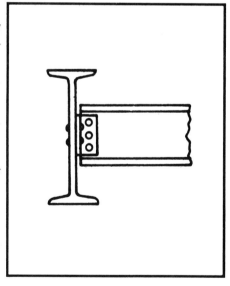

Fig. I-1: Cross-section of I beam.

IC: Pronounced "eye see": 1) Interrupting capacity of any device required to break current (switch, circuit breaker, fuse, etc.); the amount of current that the device can interrupt without damage to itself. 2) Electronics: integrated circuit.

ID: Inside diameter.

identified: Marked to be recognized as grounded.

IEC: International Electrochemical Commission.

IEEE: Institute of Electrical and Electronics Engineers.

ignition transformer: A transformer designed to provide a high voltage current.

IIR (Isobutylene Isoprene Rubber): Butyl synthetic rubber.

impedance: The opposition to current flow in an ac circuit; impedance includes resistance (R), capacitive reactance (X_C) and inductive reactance (X_L); it is measured in ohms.

impedance matching: Matching source and load impedance for optimum energy transfer with minimum distortion.

impulse: A surge of unidirectional polarity.

incandescence: Glowing due to heat.

incandescent: That which gives light or glows at a white heat.

incandescent lamp: An electric lamp or bulb containing a thin wire or filament of infusible conducting material. See Fig. I-2.

inching: Momentary activation of machinery used for inspection or maintenance. Controls for such activation usually consists of *start, stop,* and *inch.* The device used to operate such machinery is usually the ac induction motor — either single- or three-phase.

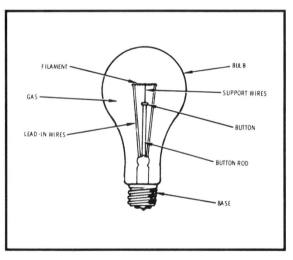

Fig. I-2: Incandescent lamp.

incombustible material: Material that will not ignite or actively support combustion in a surrounding temperature of 1200 degrees Fahrenheit during an exposure of five minutes; also, material that will not melt when the temperature of the material is maintained at 900 degrees Fahrenheit for at least five minutes.

indirect lighting: Lighting effect obtained by focusing the light against the ceiling or some other surface from which it is diffused in the area to be lighted. See Fig. I-3.

Fig. I-3: Valances provide indirect up-light that reflects off ceiling for general room lighting and down-light for drapery accent.

indoor: Not suitable for exposure to the weather.

induced voltage: A voltage set up by a varying magnetic field linked with a wire, coil, or circuit.

inductance: The creation of a voltage due to a time-varying current; the opposition to current change, causing current changes to lag behind voltage changes; units of measure: henry.

induction: The production of magnetization or electrification in a body by the mere proximity of magnetized or electrified bodies, or of an electric current in a conductor by the variation of the magnetic field in its vicinity.

induction coil: Essentially a transformer with open magnetic circuit, in which an alternating current of high voltage is induced in the secondary by means of a pulsating direct current in the primary.

induction heater: The heating of a conducting material in a varying electromagnetic field due to the material's internal losses.

induction machine: An asynchronous ac machine to change phase or frequency by converting energy — from electrical to mechanical, then from mechanical to electrical.

induction motor: An ac motor that does not run exactly in step with the alternations. Currents supplied are led through the stator coils only; the rotor is rotated by currents induced by the varying field set up by the stator coils.

inductive reactance: The opposition to the flow of an electrical current in a circuit that consists of turns of wire. The opposition is greater if the turns are wound on an iron core. The measure of resistance to the flow of alternating current through a coil.

inductivity: The capacity or power for induction.

inductor: A device having winding(s) with or without a magnetic core for creating inductance in a circuit.

infrared lamp: An electrical device that emits infrared rays — invisible rays just beyond red in the visible spectrum.

infrared radiation: Radiant energy given off by heated bodies which transmits heat and will pass through glass.

ink: The material used for legends and color coding.

in phase: The condition existing when waves pass through their maximum and minimum values of like polarity at the same instant.

instantaneous value: The value of a variable quantity at a given instant.

instrument: A device for measuring the value of the quantity under observation.

insulated: Separated from other conducting surfaces by a substance permanently offering a high resistance to the passage of energy through the substance.

insulating tape: Adhesive tape made nonconducting by being saturated with an insulating compound or manufactured from a nonconducting material; used for covering wire joints and exposed parts.

insulating transformer: A transformer that has the primary carefully insulated from the secondary. There is no electrical metallic connection between the primary and secondary.

insulation, class rating: A temperature rating descriptive of classes of insulations for which various tests are made to distinguish the materials; not related necessarily to operating temperatures.

insulation, electrical: A medium in which it is possible to maintain an electrical field with little supply of energy from additional sources; the energy required to produce the electric field is fully recoverable only in a complete vacuum (the ideal dielectric) when the field or applied voltage is removed: used to a) save space; b) enhance safety; c) improve appearance.

insulation fall-in: The filling of strand interstices, especially the inner interstices, which may contribute to connection failures.

insulation level (cable): The thickness of insulation for circuits having ground fault detectors which interrupt fault currents within: 1) 1 minute = 100% level; 2) 1 hour = 133% level; 3) over 1 hour = 173% level.

insulation, thermal: Substance used to retard or slow the flow of heat through a wall or partition.

integral: Built into or self-contained; that is, an electric heater with an integral thermostat.

integrated circuit: A circuit in which different types of devices such as resistors, capacitors, and transistors are made from a single piece of material and then connected to form a circuit.

integrated electrical system: An industrial wiring system in which:

- An orderly shutdown is required to minimize personnel hazard and equipment damage. The conditions of maintenance and supervision assure that qualified persons will service the system.

- Effective safeguards, acceptable to the authority having jurisdiction, are established and maintained.

integrator: Any device producing an output proportionate to the integral of one variable with respect to a second variable; the second is usually time.

intercalated tapes: Two or more tapes of different materials helically wound and overlapping on a cable to separate the materials.

interconnected system: Operating with two or more power systems connected through tie lines.

interface: 1) A shared boundary. 2) (nuclear power); a junction between Class 1E and other equipment of systems.

interference: Extraneous signals or power that are undesired.

interlock: A safety device to ensure that a piece of apparatus will not operate until certain conditions have been satisfied.

interpolate: To estimate an intermediate between two values in a sequence.

interpole: A small field pole placed between the main field poles and electrically connected in series with the armature of an electric rotating machine.

interrupter: A device that opens and closes a circuit at very frequent intervals.

interrupting time: The sum of the opening time and arcing time of a circuit opening device.

interstice: The space or void between assembled conductors and within the overall circumference of the assembly.

intrinsically safe: Incapable of releasing sufficient electrical or thermal energy under normal or abnormal conditions to cause ignition of a specific hazardous atmospheric mixture in its most ignitable concentration.

inverter: An item that changes dc to ac.

ion: An electrically charged atom or radical.

ionization: 1) The process or the result of any process by which a neutral atom or molecule acquires charge. A breakdown that occurs in gaseous parts of an insulation when the dielectric stress exceeds a critical value without initiating a complete breakdown of the insulation system; ionization is harmful to living tissue, and is detectable and measurable; may be evidenced by corona. An ionization smoke detector is shown in Fig. I-4 on page 93.

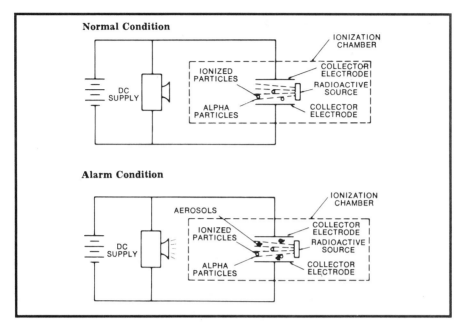

Fig. I-4: Characteristics of an ionization fire detector.

ionization factor: The difference between percent dissipation factors at two specified values of electrical stress; the lower of the two stresses is usually so selected that the effect of the ionization on dissipation factor at this stress is negligible.

IPCEA (Insulated Power Cable Engineers Association): The association of cable manufacturing engineers who make nationally recognized specifications and tests for cables.

IR (Insulation resistance): The measurement of the dc resistance of insulating material; can be either volume or surface resistivity; extremely temperature sensitive.

IR drop: The voltage drop across a resistance due to the flow of current through the resistor.

IRK (Insulation dc resistance constant): A system to classify materials according to their resistance on a 1000 foot basis at 15.5°C (60°F).

irradiation, atomic: Bombardment with a variety of subatomic particles; usually causes changes in physical properties.

ISO: International Organization for Standardization who have put together the "SI" units that are now the international standards for measuring.

isolated: Not readily accessible to persons unless special means for access are used.

isolating: Referring to switches, this means that the switch is not a loadbreak type and must only be opened when no current is flowing in the circuit. This term also refers to transformers (an isolating transformer) used to provide magnetic isolation of one circuit from another, thereby breaking a metallic conductive path between the circuits.

isotope: Atoms of a given element, each having different mass from the other because of different quantities of subatomic particles in the nucleus; isotopes are useful because several are naturally radiating and thus can become radiation sources for medical treatments or researching labs; a common isotope is Cobalt 60.

J

jack: A plug-in type terminal.

jacket: A non-metallic polymeric close-fitting protective covering over cable insulation; the cable may have one or more conductors.

jacket, conducting: An electrically conducting polymeric covering over an insulation.

jamb: Upright member forming the side of a door or window opening.

jamming: The wedging of a cable so that it can no longer be moved during installation.

jan: Joint army and navy specification.

JB: Pronounced "jay bee"; refers to any junction box, taken from the initial letters J and B.

joule: The derived SI unit for energy, work, quantity of heat; one joule equals one newton-meter or .73732 foot pounds. One joule per second equals one watt.

joule heat: Also called *joule effect,* the thermal effect that results when electrical current flows through a resistance. It is measured in watts. When a current of 1 A flows through a resistance of 1 Ω, the joule heat given off is equivalent to 1 W. This is explained by Ohm's law for power, which reads:

$$P - I^2R$$

where P is power in watts, I is current in amperes, and R is resistance in ohms. Using this formula, when 2 A of current flow through a resistance of 1 Ω, the total power dissipation is 4 W.

Joule's law: The heat generated in a conductor by an electric current is proportional to the resistance of the conductor, the time during which the current flows, and the square of the strength of the current. The quantity of heat in calories may be calculated by the use of the equation:

$$\text{Calories per second} = \text{Volts} \times \text{Ampere} \times 0.24$$

The total number of calories or heat developed in seconds will be given by:

$$\text{Heat} = \text{Volts} \times \text{Amperes} \times \text{Seconds} \times 0.24$$

journeyman: Properly, one who has gained a thorough knowledge of his or her trade by serving an apprenticeship, although the term is often applied to any worker who is sufficiently skilled to command the standard rate of journeyman's pay.

jumper: A short length of conductor, usually a temporary connection. See Fig. J-1.

junction: A connection of two or more conductors. Place of union; point of meeting; joint.

junction box: An enclosure where one or more raceways or cables enter, and in which electrical conductors can be, or are, spliced.

jurisdiction: The area assigned to a local labor union.

jute: A fibrous natural material used as a cable filler or bedding.

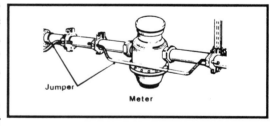

Fig. J- 1: Bonding jumper wires used to ensure continuity at a water meter. See NE Code Section 250-112.

ka: KiloAmpere.

kc: Kilocycle; use kiloHertz.

kcmil: One thousand circular mils. 250 kcmil is a conductor size of 250,000 circular mils; 500 kcmil is 500,000 circular mils, etc.

kel: Kilogram.

Kelvin (K): The basic SI unit of temperature: $\frac{1}{273.16}$ of thermodynamic temperature of the triple freezing point of water.

Kelvin double bridge: A special bridge that is used for measuring very low resistance (0.1 ohm or less). The arrangement of the bridge reduces the effects of contact resistance, which causes significant error when such low resistances are connected to conventional resistance bridges.

Kelvin's temperature: Any temperature in the absolute scale. This is a temperature scale with its zero point at -273.1°C, or absolute zero. The unit of thermodynamic temperature is the *kelvin,* and its symbol is K.

keyhole saw: A small tapered-blade saw used for cutting keyholes, small openings for fishing wires, and the like. See Fig. K-1.

kHz: kiloHertz. A unit of frequency that is equivalent to 1000 Hz, or 1000 cycles.

kilo: 10^3 or 1000.

kilocycle: One thousand cycles.

Fig. K-1: Keyhole saw.

kilogram (kg): The basic SI unit for mass; an arbitrary unit represented by an artifact kept in Paris, France.

kilometer: A metric unit of linear measurement equal to 1000 meters.

kilovolt ampere: One thousand volt-amperes.

kilowatt: Unit of electrical power equal to 1000 watts.

kilowatt-ft: The product of load in kilowatts and the circuit's distance over which a load is carried in feet; used to compute voltage drop.

kilowatt hour: The work performed by one kilowatt of electric power in one hour. The unit on which the price of electrical energy is based.

kinetic energy: Energy by virtue of motion. Whenever work is accomplished on an object, energy is consumed (changed from one kind to another). If no energy is available, no work can be performed. Thus, energy is the ability to do work. One form of energy is that which is contained by an object in motion. In driving a nail into a block of wood, a hammer is set in motion in the direction of the nail. As the hammer strikes the nail, the energy, or motion of the hammer, is converted into work as the nail is driven into the wood. This energy is called *kinetic energy*.

Kirchoff's Laws: 1) The algebraic sum of the currents at any point in a circuit is zero. That is:

$$I_1 + I_2 + I_3 + \ldots = 0$$

where I_1, I_2, I_3, etc. are the currents entering and leaving the junction. Currents entering the junction are assumed to be positive, while currents leaving the junction are negative. When solving a problem using this equation, the currents must be placed into the equation with the proper polarity signs attached. 2) The algebraic sum of the product of the current and the impedance in each conductor in a circuit is equal to the electromotive force in the circuit.

knee: The bend in a response curve that is most often an indication of the onset of saturation or cutoff. The point on the curve that represents the knee marks an abrupt change.

knife-blade fuse: A fuse having end connections that resemble the blades of a knife switch and that fit into the contacts of the cutout in the same manner that the blades of a switch fit into the switch contacts.

knife switch: A switch that opens or closes a circuit by the contact of one of more blades between two or more flat surfaces or contact blades. See Fig. K-2.

knob: A porcelain device for holding electrical conductors in place. For open wiring on insulators, the knob is used in conjunction with cleats, tubes, and flexible tubing for the protection and support of single insulated conductors run in or on buildings, and not concealed by the building structure. When concealed, this wiring method is called "concealed knob-and-tube" wiring. Open wiring on insulators has been limited

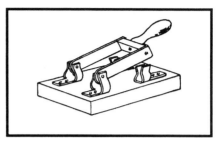

Fig. K-2: Knife switch.

to farm and industrial applications; concealed knob-and-tube wiring is allowed only as an extension of an existing system or else with special permission from the inspecting authorities. See Fig. K-3.

knockout: A portion of an enclosure designed to be readily removed for installation of a raceway or cable connector. Sometimes called "concentric" or "eccentric." See Fig. K-4 on page 100.

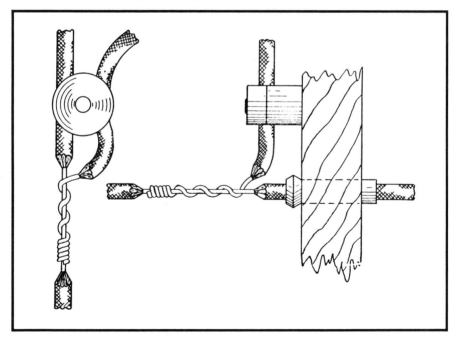

Fig. K-3: Open wiring on insulators is allowed only for industrial and farm applications.

KO: Pronounced "kay oh"; a knock-out, the partially cut opening in boxes, panel cabinets and other enclosures, which can be easily knocked out with a screw driver and hammer to provide a clean hole for connecting conduit, cable or some fittings. See *knockout.*

kVA: Kilovolts times Ampere. Also referenced as 1000 volt-amperes.

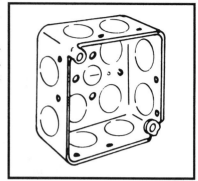

Fig. K-4: An outlet box showing knockouts.

L

LA: Lightning arrestor.

labeled: Items to which a label, trademark, or other identifying mark of nationally recognized testing labs has been attached to identify the items as having been tested and meeting appropriate standards.

lacquer: A protective coating or finish that dries to form a film by evaporation of a volatile constituent.

lagging: The wood covering for a reel.

lally column: Concrete-filled cylindrical steel structural column.

laminated core: An assembly of steel sheets for use as an element of magnetic circuit; the assembly has the property of reducing eddy-current losses.

laminated wood: Wood built up of piles or laminations that have been joined either with glue or with mechanical fasteners. The piles usually are too thick to be classified as veneer, and the grain of all piles is parallel.

lamp: A device to convert electrical energy to radiant energy, normally visible light; usually only 10-20% is converted to light. See Fig. L-1 on page 102.

lap: The relative position of applied tape edges; "closed butt lap" — tapes just touching; "open butt" or "negative lap" — tapes not touching; "positive lap" or "lap" — tapes overlapping.

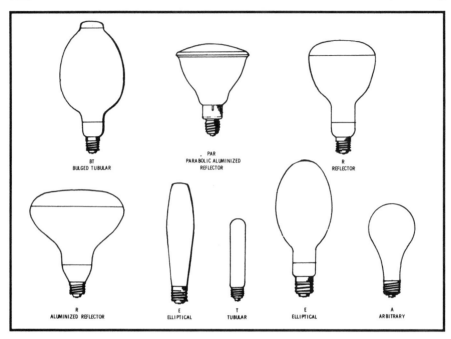

Fig. L-1: Several types of lamps in current use.

lap winding: A parallel winding that has an even number of segments and slots. The number of bars, however, is equal to or a multiple of the number of slots. It is known as a parallel winding because there are as many circuits in parallel as there are poles. Thus, a four-pole motor has four circuits in parallel, and a six-pole motor has six circuits in parallel through the armature. This requires that there be as many brushes as there are poles for the motor.

Since there are so many paths in parallel, a larger current can pass through the winding because it has several paths to travel. This, in turn, reduces the amount of voltage needed in the armature to push the current through so many paths.

The lead connections for a lap winding come from each side of the coil as top and bottom leads and connect adjacent to each other on the commutator. The position that they connect on the commutator cannot be set down to a hard-and-fast theoretical rule, but is rather at the discretion of the manufacturer who designs the motor.

latent heat: Heat given off or absorbed in a process (as vaporization or fusion) other than a change in temperature.

law of charges: Like charges repel, unlike charges attract.

law of magnetism: Like poles repel, unlike poles attract.

lay: The axial length of one turn of the helix of any element in a cable.

lay direction: Direction of helical lay when viewed from the end of the cable.

lay length: Distance along the axis for one turn of a helical element.

lead (leed): A short connecting wire brought out from a device or apparatus.

lead-acid battery: The most widely used type of storage battery, which has an EMF (voltage) of 2.2 V per cell. In its charged condition, the active materials in the lead-acid battery are lead dioxide (sometimes referred to as lead peroxide) and spongy lead. The lead dioxide is used as the positive plate, while the spongy lead forms the negative plate.

leading: Applying a lead sheath.

lead squeeze: The amount of compression of a cable by a lead sheath.

leakage: Undesirable conduction of current.

leakage distance: The shortest distance along an insulation surface between conductors.

leakage resistance: The ohmic value of the path between two electrodes that are insulated from each other.

leg: A portion of a circuit, such as a switch leg or switch loop.

legend, embossed: Molded letters and numbers in the jacket surface; letters may be raised or embedded.

Lenz' Law: "In all cases the induced current is in such a direction as to oppose the motion which generates it." That which states that whenever the value of an electric current is changed in a circuit, it creates an electromotive force by virtue of the energy stored up in its magnetic field, which opposes the change.

light: Radiant energy lying within a wavelength that spreads from 100 to 10,000 nm. The human eye can perceive light radiations in a frequency range of between 450 and 700 nm. The color of light is determined by its wavelength. Energy at the short-wave end of the visible spectrum, from 380 to about 450 nm, produces the sensation of violet. The longest visible waves, from approximately 630 to 760 nm, appear as red. Between these lie the wavelengths that the eye sees as blue (450-490 nm), green (490-560 nm), yellow (560-590 nm), and orange (590-630 nm), the colors of the rainbow.

lightning arrestor: A device designed to protect circuits and apparatus from high transient voltage by diverting the over-voltage to ground. They should be placed on upward projections such as chimneys, towers, and the like. On flat roofs, rods should be placed 50 ft. on center, and on edges of flat roofs and ridges of pitched roofs, about 25 ft. on center. The rods should project from 10 to 60 in. above flat roofs and ridges, and from 10 to 14 in. above upward projections. See *surge arrestor*.

light quantity: A measure of the total light emitted by a source of light falling on a surface. It may be expressed in lumens or watts. If all the light were of green-yellow color, at the peak of the spectral luminosity curve, 1 W would be equal to 681 lm. This gives some idea of the relative luminous efficiency of common light sources. For instance, a 100-W incandescent lamp emits about 1600 lm, only 2.4 W as light and the balance as heat. A 40-W fluorescent lamp emits about 3100 lm, 4.5 W as light and the balance as heat. From this, it can be seen that the fluorescent lamp gives four to five times as much light as the incandescent on a watt-for-watt basis.

light relay: A photoelectric device that triggers a relay in accordance with fluctuations in the intensity of a light beam. These may be purely electronic in nature or electromechanical, depending on the individual circuit. Often, light relays use an active or passive photoelectric device in the base circuit of a transistor, which is connected in series with an electromechanical relay. When light strikes the photocell, the transistor is driven to saturation and conducts current through the relay coil, causing it to change states.

light wave: A stream of electromagnetic energy that falls into the light spectrum. This lies between 100 and 10,000 nm. Visible light (that which can be seen by the human eye) lies within a very narrow band width of this spectrum at a frequency range of about 500 to 600 nm.

LIM (Laboratory Inspection Manual): A summary of specified values for quality testing.

limit: A boundary of a controlled variable.

limit control: Control used to open or close electrical circuits as temperature or pressure limits are reached.

limiter: A device in which some characteristic of the output is automatically prevented from exceeding in predetermined value.

line: A circuit between two points; ropes used during overhead construction.

linear: Arranged in a line.

linearity: When the effect is directly proportional to the cause.

line, bull: A rope for large loads.

line, finger: A rope attached to a device on a pole when a device is hung, so further conductor installation can be done from the ground.

line, pilot: A small rope strung first.

line, tag: A rope to guide devices being hoisted.

liquid absorbent: A chemical in liquid form that has the property to absorb moisture.

liquid line: The tube that carries liquid refrigerant from the condenser or liquid receiver to the refrigerant control mechanism.

liquid receiver: Cylinder connected to a condenser outlet for storage of liquid refrigerant in a system.

Lissajous Figure: A special case of an x-y plot in which the signals applied to both axes are sinusoidal functions; useful for determining phase and harmonic relationships.

listed: Items in a list published by a nationally recognized independent lab that makes periodic tests. See Fig. L-2. See *labeled.*

liter: Metric unit of volume that equals 61.03 cubic inches.

live: Energized.

live-front: Any panel or other switching and protection assembly, such as switchboard or motor control center, which has exposed electrically energized parts on its front, presenting the possibility of contact by personnel.

live load: Any load on a structure other than a dead load includes the weight of persons occupying the building and freestanding material.

Fig. L-2: Underwriters' Laboratories' label.

LMFBR (liquid metal, fast-breeder reactor) (nuclear power): A basic nuclear power fission reactor in which a metal (sodium) is heated by the reactor; this heats a second metal loop that transfers its heat to a third loop to produce steam for the turbine; this reactor can also produce other nuclear fuel at the same time.

LMP: Low Molecular Weight Polyethylene.

load: 1) A device that receives power. 2) The power delivered to such a device.

load-break: Referring to switches or other control devices, this phrase means that the device is capable of safely interrupting load current — to distinguish such devices from other disconnect devices that are not rated for breaking load current and must be opened only after the load current has been broken by some other switching device.

load center: An assembly of circuit breakers or switches. See Fig. L-3.

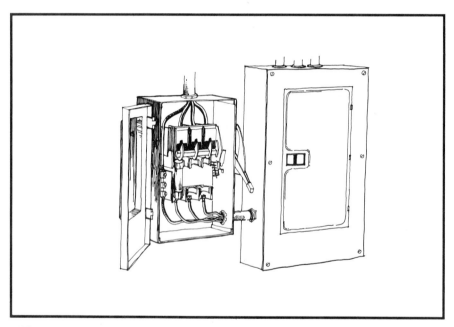

Fig. L-3: A safety switch acting as a disconnect for a load center or panelboard.

load factor: The ratio of the average to the peak load over a certain period. The time may be either the normal number of operating hours per day or 24 h, as generally used by the power companies. The average load is equal to the kilowatt-hours used in the specified time, as measured by a watthour meter, divided by the number of hours. The maximum load is the highest load at one time, as measured by some form of maximum-demand or curve-drawing watthour meter.

load losses: Those losses incidental to providing power.

LOCA (Loss of Coolant Accident) (nuclear power): The test to simulate nuclear reactor accident exhibited by high radiation, high temperature, etc.

location, damp: A location subject to a moderate amount of moisture such as some basements, barns, cold-storage, warehouses, and the like.

location, dry: A location not normally subject to dampness or wetness; a location classified as dry may be temporarily subject to dampness or wetness, as in the case of a building under construction.

location, wet: A location subject to saturation with water or other liquids.

locked rotor: The condition occurring when the circuits of a motor are energized but the rotor is not turning.

locknut: A fitting for securing cable and conduit connectors to outlet boxes, panelboards, and other apparatus; also used on threaded metal conduit. See Fig. L-4.

lockout: To keep a circuit locked open. To "tag out" a circuit as per OSHA rules.

lock seam: Joining of two sheets of metal consisting of a folded, pressed, and soldered joint.

logarithm: The exponent that indicates the power to which a number is raised to produce a given number.

logarithmic: Pertaining to the function y = log x.

longwall machine (mining): A machine used to undercut coal at relatively long working spaces.

looping-in: Avoiding splices by looping wire through device connections instead of cutting the wire.

Fig. L-4: Locknut.

loss: Power expended without doing useful work.

lug: A device for terminating a conductor to facilitate the mechanical connection. Fig. L-5 on page 108.

lumen: The derived SI unit for luminous flux.

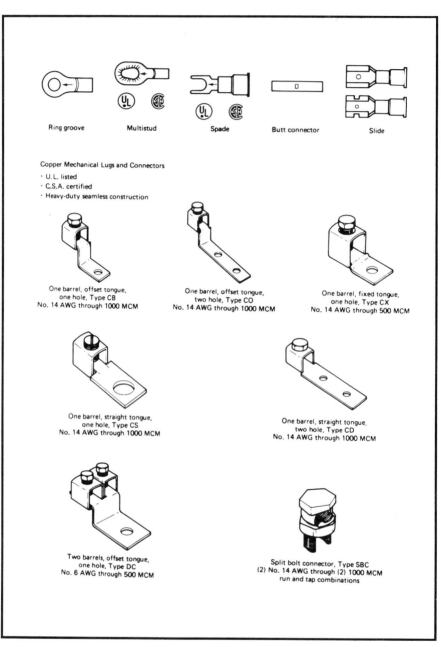

Ring groove   Multistud   Spade   Butt connector   Slide

Copper Mechanical Lugs and Connectors
· U.L. listed
· C.S.A. certified
· Heavy-duty seamless construction

One barrel, offset tongue,
one hole, Type CB
No. 14 AWG through 1000 MCM

One barrel, offset tongue,
two hole, Type CO
No. 14 AWG through 1000 MCM

One barrel, fixed tongue,
one hole, Type CX
No. 14 AWG through 500 MCM

One barrel, straight tongue,
one hole, Type CS
No. 14 AWG through 1000 MCM

One barrel, straight tongue,
two hole, Type CD
No. 14 AWG through 1000 MCM

Two barrels, offset tongue,
one hole, Type DC
No. 6 AWG through 500 MCM

Split bolt connector, Type SBC
(2) No. 14 AWG through (2) 1000 MCM
run and tap combinations

Fig. L-5: Several types of lugs used for wire connections.

luminaire: A complete lighting unit that includes the lamp, sockets, and equipment for controlling light, such as reflectors and diffusers. On electric discharge lighting, the luminaire also includes a ballast. The common term used for luminaire is *lighting fixture,* or in some cases, simple fixture.

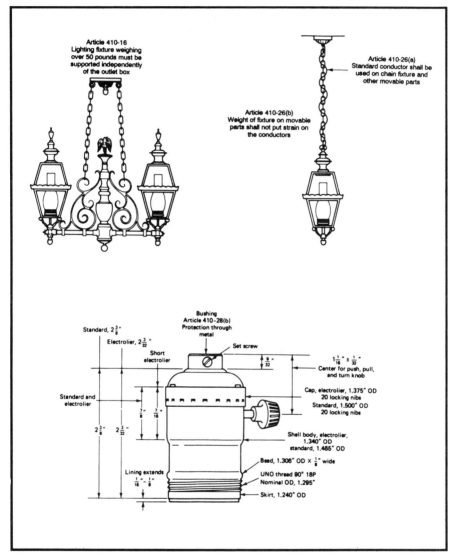

Fig. L-6: Several types of incandescent luminaires or lighting fixtures.

luminance: Intensity in a given direction divided by projected area, as intercepted by the air. It is subjective intensity and ranges from very dim to very bright. Luminance is expressed as candelas per square inch in a certain direction. Candelas per square inch may be put into more convenient form by multiplying by 452, giving luminance in footlamberts. Another way of looking at luminance is in relation to illumination and the reflection factor. For a nonspecular surface:

Luminance = Illumination x Reflection factor

or

$$L = E \; x \; R$$

where E equals footcandles, R equals reflection factor, and L equals footlamberts.

To illustrate, if $E = 100$ fc and $R = 50\%$, then $L = 100$ x $0.50 = 50$ footlamberts.

lus: The derived SI unit for illuminance.

LV: Low voltage.

LWBR (nuclear): Light water breeder reactor.

M

machine: An item to transmit and modify force or motion, normally to do work.

machine rating: The amount of power a machine can deliver without overheating.

machine screw: A very commonly used type of screw with clear-cut threads and a variety of head shapes. It may be used either with or without a nut.

magnet: A body that produces a magnetic field external to itself; magnets attract iron particles.

magnetic circuit breaker: An electromagnetic device for opening a circuit.

magnetic coil: The winding of an electromagnet. A coil of wire wound in one direction, producing a dense magnetic field capable of attracting iron or steel when carrying a current of electricity.

magnetic cutout: A device for breaking an electrical circuit by means of an electromagnet, instead of by fusing a part of the circuit.

magnetic density: The number of lines of force or induction per unit area taken perpendicular to the induction. In free space, flux density and field intensity are the same numerically, but within magnetic material, the two are quite different.

magnetic field: 1) A magnetic field is said to exist at a point if a force over and above any electrostatic force is exerted on a moving charge at the point. 2) The force field established by ac through a conductor, especially a coiled conductor.

magnetic flux: Magnetic lines of force set up by an electromagnet, permanent magnet, or solenoid.

magnetic force: The force by which attraction and repulsion are exerted by the poles of a magnet.

magnetic induction: The number of magnetic lines or the magnetic flux per unit of cross-sectional area perpendicular to the direction of the flux.

magnetic permeability: A measure of the ease with which magnetism passes through an substance.

magnetic pole: Those portions of the magnet toward which the external magnetic induction appears to converge (south) or diverge (north).

magnetic switch: A switch operated or controlled by an electromagnet.

magnetism: 1) That property of iron, steel, and some other substances, by virtue of which, according to fixed laws, they exert forces of attraction and repulsion. 2) The science that treats the conditions and laws of magnetic force.

make and break: The term may be applied to several electrical devices. Primarily, there is a pair of contact points, one stationary, and the other operated by a cam that makes the break in a circuit between these points.

manhole: A subsurface chamber, large enough for a man, to facilitate cable installation, splices, etc. in a duct bank.

manual: 1) Operated by mechanical force applied directly by personal intervention. 2) A handbook of instructions.

marker: A tape or colored thread in a cable that identifies the cable manufacturer.

mass: The property that determines the acceleration the body will have when acted upon by a given force; unit = kilogram.

master switch: A main switch; a switch controlling the operation of other switches.

mat: A concrete base for heavy electrical apparatus, such as transformers, motors, generators, etc.; sometimes the term includes the concrete base, a bed of crushed stone around it and an enclosing chain-link fence.

matrix: A multi-dimensional array of items.

matte surface: A surface from which reflection is predominately diffused.

maximum voltage: The highest voltage reached in each alternation of an alternating voltage.

MCM: An expression referring to conductors of sizes from 250 MCM, which stands for Thousand Circular Mils, up to 2000 MCM. The newest way of expressing thousand circular mils is kcmil; that is, 250 kcmil, 500 kcmil, etc.

mean: An intermediate value; arithmetic—sum of values divided by the quantity of the values; the average.

mechanical water absorption: A check of how much water will be absorbed by material in warm water for seven days (mg/sq. in. surface).

medium hard: A relative measure of conductor temper.

meg or mega: When prefixed to a unit of measurement it means one million times that unit.

megavolt: A unit of voltage equal to one million volts.

megger®: A test instrument for measuring the insulation resistance of conductors and other electrical equipment; specifically, a megohm (million ohms) meter; this is a registered trade name of the James Biddle Co. See Fig. M-1.

megohm: A unit of electrical resistance equal to one million ohms.

megohmmeter: An instrument for measuring extremely high resistance.

melt index: The extrusion rate of a material through a specified orifice at specified conditions.

melting time: That time required for an overcurrent to sever a fuse.

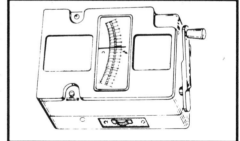

Fig. M-1: Megger manufacturedby the James Biddle Co.

mercury vapor lamp: A type of lamp that produces light by passing a current through mercury vapor.

messenger: The supporting member of an aerial cable.

metal clad (MC): The cable core is enclosed in a flexible metal covering.

metal-clad switchgear: Switchgear having each power circuit device in its own metal enclosed compartment.

metal filament: The electrical conductor that glows when heated in an incandescent lamp. See Fig. M-2.

meter: An instrument designed to measure; metric unit of linear measurement equal to 39.37 inches.

meter pan: A shallow metal enclosure with a round opening, through which a single kilowatt-hour meter is mounted, as the usual meter for measuring the amount of energy consumed by a particular building or other electrical system.

metric system: An international language of measurement. Its symbols are identical in all languages. The conversion factors in Fig. M-3 provide conversions from metric to English and from English to metric units, along with many other useful conversions.

MFT: Thousands of feet.

mho: Reciprocal of ohm.

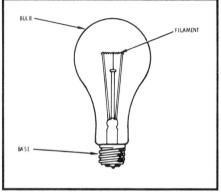

Fig. M-2: Basic components of an incandescent lamp, including the metal filament.

Conversion Factors

Multiply	by	to obtain
acres	43,560	square feet
acres	4047	square meters
acres	1.562×10^{-3}	square miles
acres	5645.38	square varas
acres	4840	square yards
amperes	0.11	abamperes

Fig. M-3: Conversion factors for metric to English, English to metric, and other units of measure.

Multiply	by	to obtain
atmospheres	76.0	cm of mercury
atmospheres	29.92	inches of merc.
atmospheres	33.90	feet of water
atmospheres	10.333	kg per sq. meter
atmospheres	14.70	pounds per sq. inch
atmospheres	1.058	tons per sq. foot
British thermal units	0.2520	kilogram-calories
British thermal units	777.5	foot-pounds
British thermal units	3.927×10^{-4}	horsepower-hours
British thermal units	1054	joules
British thermal units	107.5	kilogram-meters
British thermal units	2.928×10^{-4}	kilowatt-hours
Btu per min.	12.96	foot-pounds per sec.
Btu per min.	0.02356	horsepower
Btu per min.	0.01757	kilowatts
Btu per min.	17.57	watts
Btu per sq. ft./Min.	0.1220	watts per sq. inch
bushels	1.244	cubic feet
bushels	2150	cubic inches
bushels	0.03524	cubic meters
bushels	4	pecks
bushels	64	pints(dry)
bushels	32	quarts(dry)
centimeters	0.3397	inches
centimeters	0.01	meters
centimeters	393.7	mils
centimeters	10	millimeters
centimeter-grams	980.7	centimeter-dynes
centimeter-grams	10^{-5}	meter-kilograms
centimeter-grams	7.233×10^{-5}	pound-feet
centimeters of mercury	0.01316	atmospheres

Fig. M-3: Conversion factors (cont'd).

Multiply	by	to obtain
centimeters of mercury	0.4461	feet of water
centimeters of mercury	136.0	kg per sq. meter
centimeters of mercury	27.85	pounds per sq. meter
centimeters of mercury	0.1934	pounds per sq. inch
centimeters per second	1.969	feet per minute
centimeters per second	0.03281	feet per second
centimeters per second	0.036	kilometers per hour
centimeters per second	0.6	meters per minute
centimeters per second	0.02237	miles per hour
centimeters per second	3.728×10^{-4}	miles per minute
cubic centimeters	3.531×10^{-5}	cubic feet
cubic centimeters	6.102×10^{-2}	cubic inches
cubic centimeters	10	cubic meters
cubic centimeters	1.308×10^{-6}	cubic yards
cubic centimeters	2.642×10^{-4}	gallons
cubic centimeters	10	liters
cubic centimeters	2.113×10^{-3}	pints(liq)
cubic centimeters	1.057×10^{-3}	quarts(liq)
cubic feet	62.43	pounds of water
cubic feet	2.832×10^{4}	cubic cm
cubic feet	1728	cubic inches
cubic feet	0.02832	cubic meters
cubic feet	0.03704	cubic yards
cubic feet	7.481	gallons
cubic feet	28.32	liters
cubic feet	59.84	pints (liq)
cubic feet	29.92	quarts (liq)
cubic feet per minute	472.0	cubic cm per sec.
cubic feet per minute	0.1247	gallons per sec.
cubic feet per minute	0.4720	liters per sec.

Fig. M-3: Conversion factors (cont'd).

Multiply	by	to obtain
cubic feet per minute	62.4	lb of water per min.
cubic inches	16.39	cubic centimerters
cubic inches	5.787×10^{-4}	cubic feet
cubic inches	1.639×10^{-5}	cubic meters
cubic inches	2.143×10^{-5}	cubic yards
cubic inches	4.329×10^{-3}	gallons
cubic inches	1.639×10^{-2}	liters
cubic inches	0.03463	pints(liq)
cubic inches	0.01732	quarts(liq)
cubic yards	7.646×10^{5}	cubic centimerters
cubic yards	27	cubic feet
cubic yards	46,656	cubic inches
cubic yards	0.7646	cubic meters
cubic yards	202.0	gallons
cubic yards	764.6	liters
cubic yards	1616	pints(liq)
cubic yards	807.9	quarts(liq)
cubic yards per minute	0.45	cubic feet per sec.
cubic yards per minute	3.367	gallons per second
cubic yards per minute	12.74	liters per second
degrees (angle)	60	minutes
degrees (angle)	0.01745	radians
degrees (angle)	3600	seconds
dynes	1.020×10^{-3}	grams
dynes	7.233×10^{-5}	poundals
dynes	2.248×10^{-6}	pounds
ergs	9.486×10^{-11}	kilograms
ergs	1	dyne-centimeters
ergs	7.376×10^{-8}	foot-pounds
ergs	1.020×10^{-3}	gram-centimeters
ergs	10^{-7}	joules

Fig. M-3: Conversion factors (cont'd).

Multiply	by	to obtain
ergs	2.390×10^{-11}	kilogram-calories
ergs	1.020×10^{-8}	kilogram-meters
feet	12	inches
feet	0.3048	meters
feet	0.36	varas
feet	1/3	yards
feet of water	0.02950	atmospheres
feet of water	0.8826	inches of mercury
feet of water	304.8	kg per sq. meter
feet of water	62.43	pounds per sq. ft.
feet of water	0.4335	pounds per sq. inch
foot-pounds	1.286×10^{-3}	British thermal units
foot-pounds	1.356×10^{7}	ergs
foot-pounds	5.050×10^{-7}	horsepower hours
foot-pounds	1.356	joules
foot-pounds	3.241×10^{-4}	kilogram-calories
foot-pounds	0.1383	kilogram-meters
foot-pounds	3.766×10^{-7}	kilowatt-hours
foot-pounds per minute	1.286×10^{-3}	Btu per minute
foot-pounds per minute	0.01667	foot pounds per sec.
foot-pounds per minute	3.030×10^{-5}	horsepower
foot-pounds per minute	3.241×10^{-4}	kg-calories per min.
foot-pounds per minute	2.260×10^{-5}	kilowatts
foot-pounds per sec.	7.717×10^{-2}	Btu per minute
foot-pounds per sec.	1.818×10^{-3}	horsepower
foot-pounds per sec.	1.945×10^{-2}	kg-calories per min.
foot-pounds per sec.	1.356×10^{-3}	kilowatts
gallons	8.345	pounds of water
gallons	3785	cubic centimeters
gallons	0.1337	cubic feet
gallons	231	cubic inches

Fig. M-3: Coversion factors (cont'd).

Multiply	by	to obtain
gallons	3.785×10^{-3}	cubic meters
gallons	4.951×10^{-3}	cubic yards
gallons	3.785	liters
gallons	8	pints (liq)
gallons	4	quarts (liq)
gallons per minute	2.228×10^{-3}	cubic ft per sec.
gallons per minute	0.06308	liters per second
grains (troy)	1	grains (av)
grains (troy)	0.06480	grams
grains (troy)	0.04167	pennyweights (troy)
grams	980.7	dynes
grams	15.43	grains (troy)
grams	10^{-3}	kilograms
grams	10^{3}	milligrams
grams	0.03527	ounces
grams	0.03215	ounces (troy)
grams	0.07093	poundals
grams	2.205×10^{-3}	pounds
horsepower	42.44	Btu per min
horsepower	33,000	foot-pounds per min.
horsepower	550	foot-pounds per sec.
horsepower	1.014	horsepower (metric)
horsepower	10.70	kg-calories per min.
horsepower	0.7457	kilowatts
horsepower	7.457	watts
horsepower (boiler)	33,520	Btu per hour
horsepower (boiler)	9,804	kilowatts
horsepower-hours	2547	British thermal units
horsepower-hours	1.98×10^{6}	foot-pounds
horsepower-hours	2.684×10^{6}	joules
horsepower-hours	641.7	kilogram-calories

Fig. M-3: Conversion factors (cont'd).

Multiply	by	to obtain
horsepower-hours	2.737×10^5	kilogram-meters
horsepower-hours	0.7457	kilowatt-hours
inches	2.540	centimeters
inches	10^3	mils
inches	0.03	varas
inches of mercury	0.03342	atmospheres
inches of mercury	1.133	feet of water
inches of mercury	345.3	kg per sq meter
inches of mercury	70.73	pounds per sq ft.
inches of mercury	0.4912	pounds per sq in.
inches of water	0.07355	inches of mercury
inches of water	25.40	kg per sq. meter
inches of water	0.5781	ounces per sq in.
inches of water	5.204	pounds per sq ft.
inches of water	0.03613	pounds per sq in.
kilograms	980,665	dynes
kilograms	10^3	grams
kilograms	70.93	poundals
kilograms	2.2046	pounds
kilograms	1.102×10^{-3}	tons(short)
kilogram-calories	3.968	British thermal units
kilogram-calories	3086	foot-pounds
kilogram-calories	1.558×10^{-3}	horsepower-hours
kilogram-calories	4183	joules
kilogram-calories	426.6	kilogram-meters
kilogram-calories	1.162×10^{-3}	kilowatt-hours
kg-calories per min.	51.43	foot pounds per sec.
kg-calories per min.	0.09351	horsepower
kg-calories per min.	0.06972	kilowatts
kilometers	10^5	centimeters
kilometers	3281	feet

Fig. M-3: Conversion factors (cont'd).

Multiply	by	to obtain
kilometers	10^3	meters
kilometers	0.6214	miles
kilometers	1093.6	yards
kilowatt-hours	3415	British thermal units
kilowatt-hours	2.655×10^6	joules
kilowatt-hours	1.341	horsepower-hours
kilowatt-hours	3.6×10^6	joules
kilowatt-hours	860.5	kilogram-calories
kilowatt-hours	3.671×10^5	kilogram-meters
kilowatts	56.92	Btu per min.
kilowatts	4.425×10^4	foot-pounds per min.
kilowatts	737.6	foot-pounds per sec.
kilowatts	1.341	horsepower
kilowatts	14.34	kg-calories per min.
kilowatts	10^3	watts
log10N	2.303	logEN or ln N
logEN or lnN	0.4343	log10N
meters	100	centimeters
meters	3.2808	feet
meters	39.37	inches
meters	10^{-3}	kilometers
meters	10^3	millimeters
meters	1.0936	yards
miles	1.609×10^5	centimeters
miles	5280	feet
miles	1.6093	kilometers
miles	1760	yards
miles	1900.8	varas
miles per hour	44.70	centimeters per sec.
miles per hour	88	feet per minute
miles per hour	1.467	feet per second

Fig. M-3: Conversion factors (cont'd).

Multiply	by	to obtain
miles per hour	1.6093	kilometers per hour
miles per hour	0.8684	knots per hour
miles per hour	0.4470	M per sec.
months	30.42	days
months	730	hours
months	43,800	minutes
months	2.628×10^6	seconds
ounces	8	drams
ounces	437.5	grains
ounces	28.35	grams
ounces	0.0625	pounds
ounces per sq inch	0.0625	pounds per sq inch
pints (dry)	33.60	cubic inches
pints (liq)	28.87	cubic inches
pounds	444,823	dynes
pounds	7000	grains
pounds	453.6	grams
pounds	16	ounces
pounds	32.17	poundals
pounds of water	0.01602	cubic feet
pounds of water	27.68	cubic inches
pounds of water	0.1198	gallons
pounds of water per min.	2.669×10^{-4}	cubic feet per sec.
pounds per cubic foot	0.01602	grams per cubic cm.
pounds per cubic foot	16.02	kg per cubic meter
pounds per cubic foot	5.786×10^{-4}	pounds per cubic inch
pounds per cubic foot	5.456×10^{-9}	pounds per mil foot
pounds per square foot	0.01602	feet of water
pounds per square foot	4.882	kg per sq. meter
pounds per square foot	6.944×10^{-3}	pounds per sq. inch
pounds per square inch	0.06804	atmospheres

Fig. M-3: Conversion factors (cont'd).

Multiply	by	to obtain
pounds per square inch	.2.307	feet of water
pounds per square inch	2.036	inches of mercury
pounds per square inch	703.1	kg per sq. meter
pounds per square inch	144	pounds per sq. foot
quarts	32	fluid ounces
quarts (dry)	67.20	cubic inches
quarts (liq)	57.75	cubic inches
rods	16.5	feet
square centimeters	1.973×10^5	circular mils
square centimeters	1.076×10^{-3}	square feet
square centimeters	0.1550	square inches
square centimeters	10^{-6}	square meters
square centimeters	100	square millimeters
square feet	2.296×10^{-5}	acres
square feet	929.0	square centimeters
square feet	144	square inches
square feet	0.09290	square meters
square feet	3.587×10^{-8}	square miles
square feet	0.1296	square varas
square feet	1/9	square yards
square inches	1.273×10^6	circular mils
square inches	6.452	square centimeters
square inches	6.944×10^{-3}	square feet
square inches	10^6	square mils
square inches	645.2	square millimeters
square miles	640	acres
square miles	27.88×10^6	square feet
square miles	2.590	square kilometers
square miles	3,613,040.45	square varas
square miles	3.098×10^6	square yards
square yards	2.066×10^{-4}	acres

Fig. M-3: Conversion factors (cont'd).

Multiply	by	to obtain
square yards	9	square feet
square yards	0.8361	square meters
square yards	3.228×10^{-7}	square miles
square yards	1.1664	square varas
temp.(degs.C)+17.8	1.8	temp.(degs.F)
temp.(degs.F)-32	5/9	temp.(degs.C)
tons(long)	2240	pounds
tons(short)	2000	pounds
yards	0.9144	meters

Fig. M-3: Conversion factors (cont'd).

mica: A silicate which separates into layers and has high insulation resistance, dielectric strength, and heat resistance.

MI cable: Mineral insulated, metal sheathed cable.

microfarad: A unit of capacity, being one millionth of a farad.

micrometer (mike): A tool for measuring linear dimensions accurately to 0.001 inch or to 0.01 mm.

micro structure: The structure of polished and etched metals as revealed by a microscope at a magnification of more than ten diameters.

microwave: Radio waves of frequencies above one gigahertz.

mil: A unit used in measuring the diameter of wire, equal to 0.001 inch (25.4 micrometers).

MIL: Military specification.

milliampere: One thousandth of an ampere.

millivolt: One thousandth of a volt; .0001 volt.

mil scale: The heavy oxide layer formed during hot fabrication or heat treatment of metals.

minimum at a point: Specifications permit the thickness at one point to be less than the average.

minimum average: The specified average insulation or jacket thickness.

mks: Meter, kilogram, second.

mm: Millimeter; 1 meter/1000.

modem: Equipment that connects data transmitting/receiving equipment to telephone lines; a word contraction of "modulator-demodulator."

modulation: 1) The varying of a "carrier" wave by a characteristic of a second "modulating" wave. 2) To regulate by or adjust to a certain measure or proportion.

modulus of electricity: The ratio of stress (force) to strain (deformation) in a material that is elastically deformed.

mogul: A socket or receptacle (lampholder) used with large incandescent lamps of 300 watts or more.

moisture-repellent: So constructed or treated that moisture will not penetrate.

moisture-resistance: So constructed or treated that moisture will not readily injure.

molded case breaker: A circuit breaker enclosed in an insulating housing.

mole (mol): The basic SI unit for amount of substance; one mole is the amount of substance of a system that contains as many elementary entities as there are atoms in 0.012 kilograms of carbon 12.

molecule: The group of atoms that constitutes the smallest particle in which a compound or material can exist separately.

motor: An apparatus to convert from electrical to mechanical energy.

motor, capacitor: A single-phase induction motor with an auxiliary starting winding connected in series with a condenser for better starting characteristics.

motor control: Device to start and/or stop a motor at certain temperature or pressure conditions. See Fig. M-4 on page 126.

motor control center: A grouping of motor controls such as starters.

motor effect: Movement of adjacent conductors by magnetic forces due to currents in the conductors.

mouse: Any weighted line used for dropping down between finished walls to attach to cable to pull the cable up; a type of vertical fishing between walls.

MPT: Male pipe thread.

MPX: Multiplexer.

MTW: Machine tool wire.

multioutlet assembly: A type of surface or flush raceway designed to hold conductors and attachment plug receptacles.

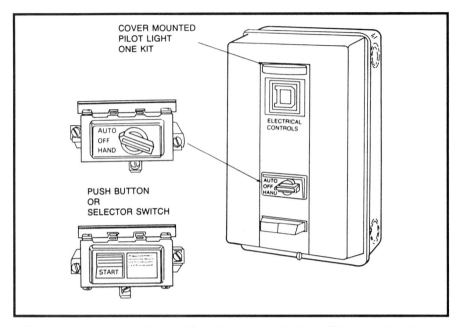

Fig. M-4: Motor controls are offered in many variations. This starter has the option of either a push button or selector switch.

multiple barrier protection (nuclear power): The keeping of radioactive fission products from the public by placing multiple barriers around the reactor; for BWR these are fuel cladding, the reactor vessel, and the containment building.

multiplex: To interleave or simultaneously transmit two or more messages on a single channel.

multiplier: A known resistance used with a multimeter or a galvanometer to increase their range. See Fig. M-5.

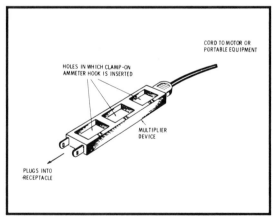

Fig. M-5: A multiplier used on an ammeter.

multispeed motor: A motor capable of being driven at any one of two or
more different speeds independent of the load. See Fig. M-6.

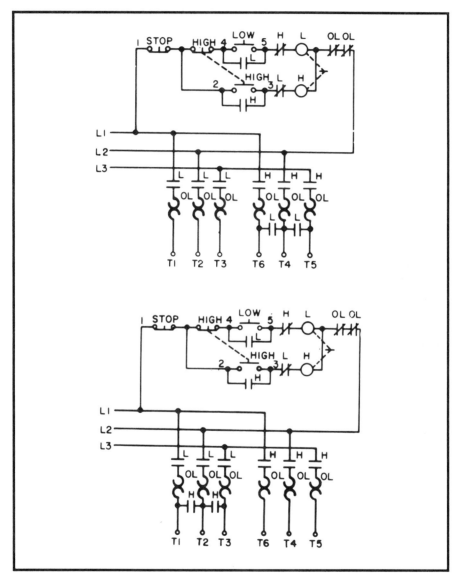

Fig. M-6: Wiring diagram of one type of multispeed motor.

mutual inductance: The condition of voltage in a second conductor because of a change in current in another adjacent conductor.

mw: Megawatt: 106 watts.

Mylar®: DuPont trade name for a polyester film whose generic name is oriented polyethylene terephthalate; used for insulation, binding tapes.

N

N/A: 1) Not available. 2) Not applicable.

National Electrical Code (NEC): A set of rules governing the selection of materials, quality of workmanship, and precautions for safety in the installation of electrical wiring. The NEC, originally prepared in 1897, is frequently revised to meet changing conditions, such as improved equipment and materials and new fire hazards. The code is the result of the best efforts of electrical engineers, manufacturers of electrical equipment, insurance underwriters, fire fighters, and other concerned experts through the country.

The NEC is published by the NFPA (National Fire Protection Association), Batterymarch Park, Quincy, MA. It contains specific rules and regulations intended to help in "the practical safeguarding of persons and property from hazards arising from the use of electricity." The NEC contains provisions considered necessary for safety. Compliance therewith and proper maintenance will result in an installation essentially free from hazard, but not necessarily efficient, convenient, or adequate for good service for future expansion of electrical use.

natural convection: Movement of a fluid or air caused by temperature change.

NBR: Nitrite-butadiene rubber: synthetic rubber.

NBS: National Bureau of Standards.

NC: Normally closed.

negative: Connected to the negative terminal of a power supply.

negative conductor: A conductor leading from the negative terminal.

negative plate: 1) In a storage cell, the spongy lead plate which, during discharge, is the negative plate or terminal. 2) In a primary cell, the carbon, copper, platinum, etc. is the negative electrode.

negative side of circuit: The conducting path of a circuit from the current-consuming device back to the source of supply.

NEMA: National Electrical Manufacturers Association.

neoprene: An oil resistant synthetic rubber used for jackets; originally a DuPont trade name, now a generic term for polychloroprene.

network: An aggregation of interconnected conductors consisting of feeders, mains, and services.

network limiter: A current limiting fuse for protecting a single conductor.

neutral: The element of a circuit from which other voltages are referenced with respect to magnitude and time displacement in steady state conditions.

neutral block: The neutral terminal block in a panelboard, meter enclosure, gutter, or other enclosure in which circuit conductors are terminated or subdivided. See Fig. N-1.

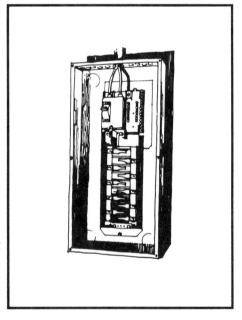

neutral wire: A grounded conductor in an electrical system that does not carry current until the system is unbalanced. Neutral conductors must have sufficient capacity for the current that they might have to carry under certain conditions. In a single-phase, three-wire system, the middle leg or neutral wire carries no current when the loads on each side between neutral are equal or balanced. If, however, the loads on the outside wires become unequal, the difference in current flows over the neutral wire.

When a 120/240-V single-phase service is used, it is highly desirable for the 120-V loads to

Fig. N-1: Interior of panelboard showing neutral block.

be balanced across both sides of the service. The neutral wire can then be smaller than the two hot or ungrounded wires. The NE Code permits the reduction of the neutral to the size that will carry the maximum unbalanced load between the neutral and any one ungrounded conductor.

A general rule of thumb is to reduce the neutral by not more than two standard wire sizes. A further demand factor of 70 percent may be applied in reducing the neutral wire for that portion of the unbalanced load that is in excess of 200 A. However, if 50 percent or more of the load consists of electric-discharge lamp ballasts, the neutral will be the same size as the ungrounded conductors.

In a three-phase, four-wire system, a three-wire branch circuit consisting of two phase wires and one neutral wire has a neutral or grounded wire that carries approximately the same current as the phase conductors. Therefore, it should be the same size as the phase conductors. However, three-phase, four-wire systems generally supply a mixed load of lamps, motors, and other appliances. Motors and similar three-phase loads connected only to the phase wires cannot throw any load onto the neutral, and such three-phase loads can be disregarded in calculating the necessary capacity for the neutral conductor.

neutron: Subatomic particle contained in the nucleus of an atom: electrical neutral. The neutron is located in the central mass, or nucleus, of the atom and possesses no electrical charge. Due to this, the neutron is not deflected by magnetic or electric fields, and its interaction with matter is mainly by collision.

newton: A unit of measurement that expresses the force of attraction or repulsion between two charged bodies. This force varies directly with the product of the individual charges, and at the same time, varies inversely with the square of the distance between charges.

NFPA (National Fire Protection Association): An organization to promote the science and improve the methods of fire protection that sponsors various codes, including the National Electrical Code.

nickel plating: The depositing of a coating of nickel on a metallic surface. Accomplished by immersion in a nickel salt bath through which an electric current of low voltage is passed.

nineteen hundred box: A commonly used term to refer to any 2-gang 4-inch square outlet box used for two wiring devices or for one wiring device with a single-gang cover where the number of wires requires this box capacity. See Fig. N-2 on page 132.

nipple: A threaded pipe or conduit of less than two feet length.

NO: Normally open.

node: A junction of two or more branches of a network.

no-load current: The current that flows through a device or circuit when the same device or circuit is delivering zero-output current. For example, when the primary of a transformer is connected to an ac source, a voltage appears at the secondary winding even when it is not connected to a load. While there is no current flow from the output in this condition, a small amount of current flows in the primary winding to sustain transformer operation. Here, the no-load current is the measured drain from the primary input voltage source. See Fig. N-3.

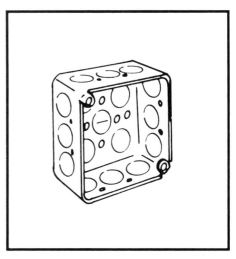

Fig. N-2: A 4-inch square or nineteen hundred box.

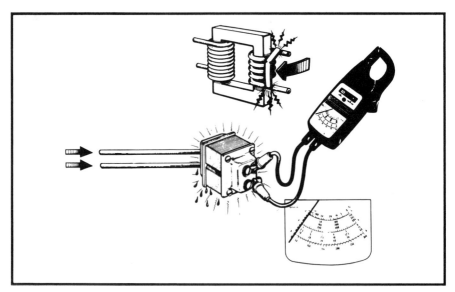

Fig. N-3: Even with the load disconnected from this transformer, there is still some current drain in the primary winding.

nominal: Relating to a designated size that may vary from the actual.

nominal rating: The maximum constant load that may be increased for a specified amount for two hours without exceeding temperature limits specified from the previous steady state temperature conditions; usually 25 or 50 percent increase is used.

nomograph: A chart or diagram with which equations can be solved graphically by placing a straightedge on two known values and reading where the straightedge crosses the scale of the unknown value.

nonautomatic: Used to describe an action requiring personal intervention for its control.

noncode installation: A system installed where there are no local, state, or national codes in force.

nonconductor: Any substance that does not allow electricity to pass through it.

noninductive circuit: A circuit in which the magnetic effect of the current flowing has been reduced by one of several methods to a minimum or to zero.

noninductive resistance: Resistance free from self-induction.

noninductive winding: A winding so arranged that the magnetic field set up by the current flowing in one half of the coil is neutralized by the magnetic field set up by the current flowing in the opposite directions in the second half.

nonmetallic-sheathed (NM) cable: A type of cable that is popular for use in residential and small commercial wiring systems. In general, it may be used for both exposed and concealed work in normally dry locations.

Type NM cable must not be installed where exposed to corrosive fumes or vapors, nor embedded in masonry, concrete, fill, or plaster, nor run in shallow chases in masonry or concrete and finished or covered with plaster or similar finish. This cable must not be used as a service-entrance cable, in commercial garages, theaters and assembly halls, motion picture studios, storage battery rooms, hoistways, hazardous locations, or embedded in poured cement, concrete, or aggregate.

For use in wood structures, holes are bored through wood studs and joists, and the cable is then pulled through these holes to the various outlets. The holes normally give sufficient support, providing they are not over 4 feet on center. When no stud or joist support is available, staples or some similar supports are required for the cable. The supports must not exceed 4.5 feet and must be within 12 inches of each outlet box or other termination point. See *Romex*.

normal charge: The thermal element charge that is part liquid and part gas under all operating conditions.

NPT: National tapered pipe thread.

NR: 1) Nonreturnable reel; a reel designed for one-time use only. 2) Natural rubber.

NRC: (Nuclear Regulatory Commission): The Federal agency for atomic element usage; formerly AEC.

NSD (neutral supported drop): A type of service cable.

Nylon® This is the DuPont trade name for polyhexamethylene-adipamide which is the thermoplastic used as insulation and jacketing material.

O

OC: Overcurrent.

OD: Outside diameter.

OF: Oxygen free.

offgassing: Percentage of a gas released during combustion.

ohm: The derived SI unit for electrical resistance or impedance; one ohm equals one volt per am-pere.

ohmmeter: An instrument for measuring resistance in ohms. An ohm-mmeter being used to trouble-shoot a small control transformer is shown in Fig. O-1 on page 136. The ohmmeter's pointer deflection is controlled by the amount of battery current passing through the moving coil. Before measuring the resistance of an unknown resistor or electrical circuit, the ohmmeter must first be calibrated. If the value of resistance to be measured can be estimated within reasonable limits, a range is selected that will give approximately half-scale deflection when the resistance is inserted between the probes. If the resistance is unknown, the selector switch is set on the highest scale. Whatever range is selected, the meter must be calibrated to read zero before the unknown resistance is measured.

ohm resistance: A dc circuit is said to have a resistance of one ohm when one volt will produce a current of one ampere through it.

Ohm's law: Mathematical relationship between voltage, current, and re-sistance in an electric circuit. Ohm's law states that the current flowing in a circuit is proportional to the voltage and inversely proportional to the resistance or opposition. See Fig. O-2 on page 137.

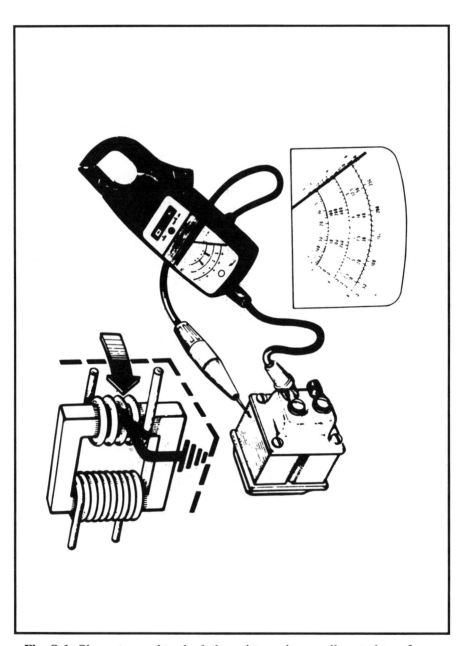

Fig. O-1: Ohmmeter used to check the resistance in a small control transformer.

OI (Official Inter-pretation): A now obsolete term used for an interpretation of the National Electrical Code made to help resolve a specific problem between an inspector and an installer. *Formal Interpretations* is the term now used to promote uniformity of interpretation and application of the provisions in the National Electrical Code. The procedures may be found in the "NFPA Regulations Governing Committee Projects."

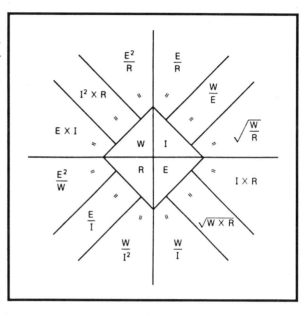

Fig. O-2: Summary of Ohm's law.

oil-: The prefix designating the operation of a device submerged in oil to cool or quench or insulate.

oil-proof: A device or appartus designed so that the accumulation of oil or vapors will not prevent safe, successful operation.

oil-tight: Construction preventing the entrance of oil or vapors not under pressure.

OL (nuclear): Operating License.

OL: Overload.

open circuit: A circuit that is energized but not allowing useful current to flow. For example, a switch installed in an energized circuit can be opened so as to stop the flow of current through the circuit.

opening time: The period between which an activation signal is initiated and switch contacts part.

open wiring: Exposed electrical conductors.

optimization: The procedure used in the design of a system to maximize or minimize some performance index.

organic: Matter originating in plant or animal life, or composed of chemicals of that origin.

orifice: Accurate size opening for controlling fluid flow.

oscillation: The variation, usually with time, of the magnitude of a quantity that is alternately greater and smaller than a reference.

oscillator: A device that produces an alternating or pulsating current or voltage electronically.

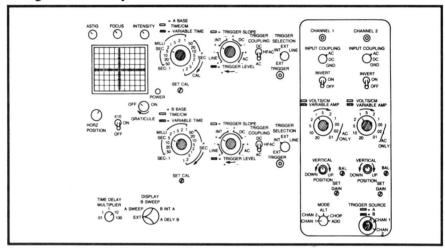

Fig. O-3: Oscilloscopes are used to show rapidly varying electrical quantities.

oscillograph: An instrument primarily for producing a graph of rapidly varying electrical quantities.

oscilloscope: An instrument primarily for making visible rapidly varying electrical quantities; oscilloscopes function similarly to TV sets. The principle components of a basic oscilloscope include a cathode-ray tube (CRT), a sweep generator, deflection amplifiers (horizontal and vertical), a power supply, and suitable controls, switches, and input connectors for proper operation. Fig. O-3 shows an oscilloscope.

OSHA (Occupational Safety and Health Act): Federal Law #91-596 of 1970 charging all employers engaged in business affecting interstate commerce to be responsible for providing a safe working place; it is administered by the Department of Labor; the OSHA regulations are published in Title 29, Chapter XVII, Part 1910 of the CFR and the Federal Register.

osmosis: The diffusion of fluids through membranes.

ought sizes: An expression referring to conductors of sizes No. 1/0, 2/0, 3/0, or 4/0. Sometimes called "naught."

outage: The condition resulting when a component is not available to perform its intended function.

outdoor: Designed for use out-of-doors.

outgassing: Dissipation of gas from a material.

outlet: A point on the wiring system at which current is taken to supply utilization equipment.

outlet box: A box installed in an electrical wiring system, from which current is taken to supply some apparatus or device.

outline lighting: An arrangement of incandescent lamps or gaseous tubes to outline and call attention to certain features, such as the shape of a building or the decoration of a window.

output: 1) The energy delivered by a circuit or device. 2) The terminals for such delivery.

oven: An enclosure and associated sensors and heaters for maintaining components at a controlled and usually constant temperature.

overcurrent protection: De-energizing a circuit whenever the current exceeds a predetermined value; the usual devices are fuses, circuit breakers, or magnetic relays.

overhead cost: The cost of operating a business, as for rent, interest on investment, maintenance and depreciation of equipment, etc.; over and above the actual cost of labor and material.

overload: Load greater than the load for which the system or mechanism was intended. A fault, such as a short circuit or ground fault, is not an overload.

overload switch: An automatic switch for breaking a circuit in case of an overload.

over-time: Time spent in working beyond specified working hours.

overvoltage (cable): Voltages above normal operating voltages, usually due to: a) switching loads on/off; b) lighting; c) single phasing.

oxidize: 1) To combine with oxygen. 2) To remove one or more electrons. 3) To dehydrogenate.

oxygen bomb test: Aging a rubber sample in a pressure container in a pure oxygen environment not deteriorated by ozone.

oxygen index: A test to rate flammability of materials in a mixture of oxygen and nitrogen.

ozone: An active molecule of oxygen that may attack insulation; produced by corona in air.

ozone resistance tests: A test for rubber, not used now because the synthetic rubbers do not deteriorate from ozone.

P

P102: A cable legend signifying acceptance for listing by Pennsylvania Department of Mines (POM).

pad: A coil of tape.

pad-mounted: A shortened expression for "pad-mount transformer," which is a completely enclosed transformer mounted outdoors on a concrete pad, without need for a surrounding chain-link fence around the metal, box-like transformer enclosure. See Fig. P-1.

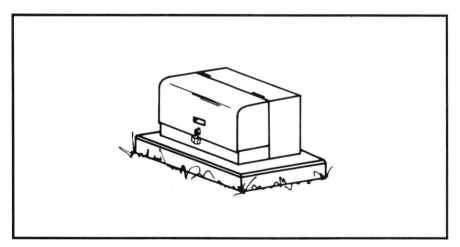

Fig. P-1. Typical pad-mounted transformer.

pan: A sheet metal enclosure for a watt-hour meter, commonly called a "meter pan."

panel: A unit for one or more sections of flat material suitable for mounting electrical devices.

panelboard: A single panel or group of panel units designed for assembly in the form of a single panel; includes buses and may come with or without switches and/or automatic over-current protective devices for the control of light, heat, or power circuits of individual as well as aggregate capacity. It is designed to be placed in a cabinet or cutout box that is in or against a wall or partition and is accessible only from the front. See Fig. P-2.

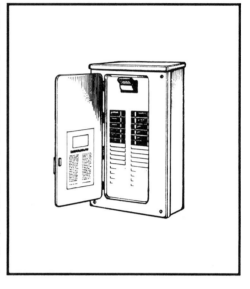

Fig. P-2: Typical electrical panelboard.

paper-lead cable: One having oil impregnated paper insulation and a lead sheath.

parallax: This occurs when the scale of an electrical instrument is not in correct alignment.

parallel: Lying side by side; extending in the same direction and equidistant at all points.

parallel circuit: A circuit having a common return between which two or more pieces of apparatus are connected, each receiving a separate portion ofthe current flow from the common feed. See Fig. P-3 on page 143.

parallel-connected transformer: Two or more transformers having their primary windings connected to the same source of supply in such a manner that the impressed voltage in each case is the same as that of the line.

parameter: A variable given a constant value for a specific process or purpose.

pascal: The derived SI unit for pressure or stress: one pascal equals one newton per square meter.

patch: To connect circuits together temporarily.

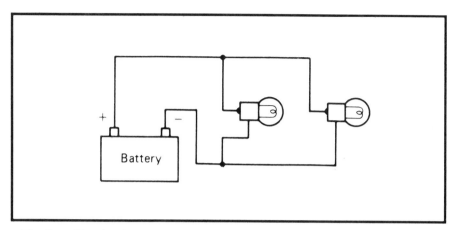

Fig. P-3: Circuit with lamps connected in parallel.

payoff: The equipment to guide the feeding of wire.

PB: Pushbutton; pull box.

PE: 1) Polyethylene. 2) Professional engineer.

peak load: The heaviest load that a generator or system is called on to supply at regular intervals, as once every 24 hours.

peak value: The largest instantaneous value of a variable.

penciling: The tapering of insulation to relieve electrical stress at a splice or termination.

penetration (nuclear): A fitting to seal pipes or cables through the containment vessel wall.

pentode: An electron tube with five electrodes or elements.

period: The minimum interval during which the same characteristics of a periodic phenomenon recur.

permalloy: An alloy of nickel and iron that is easily magnetized and demagnetized.

permanent magnet: A piece of magnet steel that retains the acquired property of attracting other pieces of magnetic material after being under the influence of a magnetic field.

permanent magnet moving-coil meter: The basic movement used in most measuring instruments for servicing electrical equipment. The basic movement consists of a stationary permanent magnet and a movable coil. When current flows through the coil, the resulting magnetic field reacts with the magnetic field of the permanent magnet and causes the coil to rotate. The greater the intensity of the current flow through he coil, the stronger the magnetic field produced; and the stronger the magnetic field, the greater the rotation of the coil.

permeability: The ability of a core of magnetic material to conduct lines of force. Some core materials have higher permeabilities than others and thus, used with a given winding, they provide inductors having greater inductance values. More precisely, the permeability of a certain magnetic material is the ratio of the flux produced with that magnetic material as the core to the flux produced with air as the core.

It is important to be aware of the fact that the permeability of a given magnetic material varies with the magnetizing force in ampere-turns per unit of core length that is applied to the core. The permeability also depends on the amount and direction of any magnetic flux that might already exist.

permissible mine equipment: Equipment that has been formally accepted by the Federal agency.

per-unit quantity: The ratio of the actual value to an arbitrary base value of a quantity.

PES: Power Engineering Society of IEEE.

PF: Power factor.

pH: An expression of the degree of acidity or alkalinity of a substance; on the scale of 1-10, acid is under 7, neutral is 7, alkaline is over 7.

phase: The fractional part "t/p" of the period through which a quantity has advanced, relative to an arbitrary origin. The time instant when the maximu, zero, or other relative value is attained by an electric wave.

phase angle: The measure of the progression of a periodic wave in time or space from a chosen instant or position.

phase conductor: The conductors other than the neutral.

phase converter: A device that will permit the operation of a three-phase induction motor from a single-phase power source. Since most commercial suppliers of power place limits on the size of single-phase motors they can serve, such limitations may also apply to three-phase motors supplied through a phase converter. The power supplier should be consulted regarding the size of converter that can be operated in a particular location.

Application of phase converters to easy-starting loads usually involves no particular problems. Such loads would include large fans and centrifugal and turbine-type irrigation pumps. For equipment that requires high starting torque or is subject to wide load fluctuation, application of a phase converter should be made only after consulting the manufacturer of the three-phase motor. Such equipment would include compressors, pumps, and barn cleaners that start under load and feed grinders and blowers on which load may vary due to uneven feeding.

Phase converters may be divided into two general types—the static converter and the rotating transformer converter. Static converters are subdivided into several types, among which are autotransformer converters, series-winding converters, and multimotor converters. Some of these must be matched in horsepower rating to the motor to be driven; that is, a 5-hp converter for a 5-hp motor, etc. The multimotor converter, as its name implies, will operate two or more motors.

The rotating transformer-type converter should have a horsepower rating as large as that of the largest motor to be driven. Additional smaller motors may be supplied by the same converter.

The starting current of a converter (three-phase motor combination) is likely to be less than for a comparable single-phase motor. By the same token, the starting torque of the motor-converter combination is likely to be less than for a similar-sized single-phase motor or for a three-phase motor operated from a three-phase supply. Likewise, the motor connected through a converter may have very little short-time overload capacity.

Overcurrent protection may be difficult to provide because of the longer starting period and the unbalanced currents that occur in the motor windings under overload conditions. Power factor is likely to be near 100 percent at rated load and slightly leading when idling on the line.

phase leg: One of the phase conductors (an ungrounded or "hot" conductor) of a polyphase electrical system.

phase meter: A meter that indicates the frequency of the circuit to which it is attached; a frequency meter.

phase out: A procedure by which the individual phases of a polyphase circuit or system are identified; such as to "phase out" a 3-phase circuit for a motor in order to identify phase A, phase B, and phase C to know how to connect them to the motor to get the correct phase rotation so the motor will rotate in the desired direction.

phase protection: A means of preventing damage in an electric motor through overheating in the event a fuse blows or a wire breaks when the motor is running. This is necessary because a motor will continue to operate on single phase even if some part of the device is damaged. In order to provide the protection, phase-failure and -reversal relays are used.

Phase-failure relays are available in a number of designs, some of which are quite complex devices.

One type of relay utilizes coils that are connected in two lines of the three-phase supply. The currents in the coils cause a rotating magnetic field to be set up to turn a copper disc clockwise. This movement, which is known as *torque*, results from one polyphase torque turning the disc clockwise and one single-phase torque turning the disc counterclockwise. The disc is kept from turning by a projection resting against a stop.

In the event that the disc rotates counterclockwise, the projecting arm causes a toggle mechanism to open line contacts, thus removing the motor from the line and preventing any damage. In the event that one line opens, the polyphase torque disappears, the single-phase torque rotates the disc counter-clockwise, and the motor is again removed from the line, thus preventing overheating.

Some complex types of phase-failure relays are commonly used in situations where it is necessary to protect not only a device, but also any persons or machines from the dangers involved in open-phase or reversed-phase sequence conditions. In some cases, a phase-failure relay may consist of a static, current-sensitive network that is connected in series with the line and a switching relay connected in the coil circuit of the starter. This sensing network services to monitor the line current in the motor. In the event one phase opens, the sensing twork detects it, causing the relay to open the starter coil current, which disconnects the motor from the line. This type of phase-failure relay usually contains a built-in delay of five cycles, which prevents nuisance drops that are caused by transient line fluctuations.

phase sequence: The order in which the successive members of a periodic wave set reach their positive maximum values; a) zero phase sequence — no phase shift; b) plus/minus phase sequence — normal phase shift.

phase shift: The absolute magnitude of the difference between two phase angles.

phasor quantity: A complex algebraic expression for sinusoidal wave.

photocell: A device in which the current-voltage characteristic is a function of incident radiation (light).

photoelectric control: A control sensitive to incident light. See Fig. P-4 on page 147.

photoelectricity: A physical action wherein an electrical flow is generated by light waves.

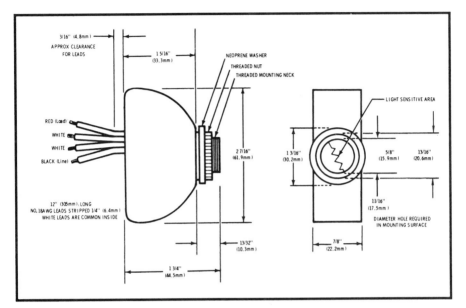

Fig. P-4: Data for a photoelectric switch.

photometer: An instrument for measuring the intensity of light or for comparing the relative intensity of different lights.

photon: An elementary quantity of radiant energy (quantum).

pi (π): The ratio of the circumference of a circle to its diameter.

pick: The grouping or band of parallel threads in a braid.

pickle: A solution or process to loosen or remove corrosion products from a metal.

pickup value: The minimum input that will cause a device to complete a designated action.

picocoulomb: 10^{-12} coulombs.

Piezoelectric effect: Some materials become electrically polarized when they are mechanically strained; the direction and magnitude of the polarization depends upon the nature, amount and the direction of the strain; in such materials the reverse is also true in that a strain results from the application of an electric field.

pigtail: A flexible conductor attached to an apparatus for connection to a circuit.

PILC cable: Paper insulated, lead covered.

pilot lamp: A lamp that indicates the condition of an associated circuit.

pilot wire: An auxiliary insulated conductor in a power cable used for control or data.

pin gauge: An insulation thickness gauge having a pin that will fit the groove left by a wire strand.

pipe vise: Type of portable vise for securing conduit during cutting, reaming, and threading. See Fig. P-5.

pitch diameter: The diameter through the center of a layer in a concentric layup of a cable or strand.

pitting: Small cavities in a metal surface.

plasma: A gas made up of charged particles.

plastic: Pliable and capable of being shaped by pressure.

plastic deformation: Permanent change in dimensions of an object under load.

Fig. P-5: Pipe vise.

plasticizer: Agent added to plastic to improve flow and processability and to reduce brittleness.

plate: A rectangular product having thickness of 0.25 inch or more.

plating: Forming an adherent layer of metal on an object.

plenum: Chamber or space forming a part of an air conditioning system.

plowing: Burying cable in a split in the earth made by a blade.

plug: A male connector for insertion into an outlet or jack.

plug fuse: A type of fuse that is held in position by a screw-thread contact instead of spring clips, as is the case with a cartridge fuse.

plugging: Braking an induction motor by reversing the phase sequence of the power to the motor. This reversal causes the motor to develop a counter torque, which results in the exertion of a retarding force. Plugging is used to secure both rapid stop and quick reversal.

Because it is possible for motor connections to be reversed when the motor is running, control circuits should be designed specifically to prevent this from occurring when it is undesirable. However, there are a number of factors that must be considered and investigated thoroughly when it is desired to have this type of operation. It may be necessary to have methods of limiting maximum permissible currents, particularly in situations with repeated operations and also with dc motors. The machine under consideration should be carefully investigated in order to ensure that this type of action will not do damage over an extended period of time.

polarity: 1) Distinguishing one conductor or terminal from another. 2) Identifying how devices are to be connected, such as positive (+) or (-).

Polarization Index: Ratio of insulation resistance measured after 10 minutes to the measure at 1 minute with voltage continuously applied.

pole: 1) That portion of a device associated exclusively with one electrically separated conducting path of the main circuit of device. 2) A supporting circular column.

poly: Polyethylene.

polychloroprene: Generic name for neoprene.

polycrystalline: Pertaining to a solid having many crystals.

polyethylene: A thermoplastic insulation having excellent electrical properties, good chemical resistance (useful as jacketing), are good mechanical properties with the exception of temperature rating.

polymer: A high-molecular-weight compound whose structure can usually be represented by a repeated small unit.

polyphase circuits: ac circuits having two or more interrelated voltages, usually of equal amplitudes, periods, phase differences, and etc.; if a neutral conductor exists, the voltages referenced to the neutral are equal in amplitude and phase; the most common version is that of three-phase, equal in amplitude with phases 120° apart. See Fig. P-6.

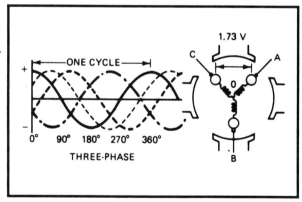

Fig. P-6: Three-phase cycle.

polyphase motor: An ac motor that is designed for either three- or two-phase operation. The two types are alike in construction, but the internal connections of the coils are different. Three-phase motors vary from fractional horsepower size to several thousand horsepower. These motors have a fairly constant speed characteristic and are made in designs giving a variety of torque characteristics. Some have a high starting torque; others have a low starting torque. Some are designed to draw a normal starting current; others draw a high starting current. They are made for practically every standard voltage and frequency and are very often dual-voltage motors. Three-phase motors are used to drive machine tools, pumps, elevators, fans, cranes, hoists, blowers, and many other machines.

Two-phase motors are like three-phase motors in all respects, except for the number of groups and the connections of the groups. As in the three-phase motor, the number of groups is equal to the number of phases multiplied by the number of poles.

polypropylene: A thermoplastic insulation similar to polyethylene, but with slightly better properties.

polytetrafluoroethylene (PTFE): A thermally stable (-90 to + 250°C) insulation having good electrical and physical properties even at high frequencies.

porcelain: Ceramic china-like coating applied to steel surfaces.

portable: Designed to be movable from one place to another, not necessarily while in operation.

positive: Connected to the positive terminal of a power supply.

positive plate: The plate of a storage cell from which the current flows to the negative plate during the process of discharging.

potential: The difference in voltage between two points of a circuit. Frequently, one is assumed to be ground (zero potential).

potential energy: Energy of a body or system with respect to the position of the body or the arrangement of the particles of the system.

potentiometer: An instrument for measuring an unknown voltage or potential difference by balancing it, wholly or in part, by a known potential difference produced by the flow of known currents in a network of circuits of known electrical constants.

pothead: A terminator for high-voltage circuit conductor to keep moisture out of the insulation and to protect the cable end, along with providing a suitable stress relief cone for shielded-type conductors. See Fig. P-7 on page 151.

power: 1) Work per unit of time. 2) The time rate of transferring energy; as an adjective, the word "power" is descriptive of the energy used to perform useful work; measurements: pound-feet per second, watts.

power, active: In a 3-phase symmetrical circuit: $p = \sqrt{3}$ VA cos ø; in a 1-phase, 2 wire circuit: $p = $ VA cos ø.

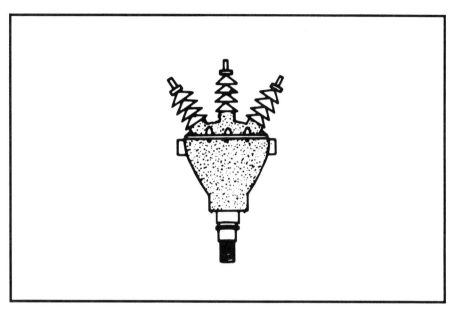

Fig. P-7: Pothead used on power pole to make transition from overhead to underground lines.

power, apparent: The product of rms volts times rms amperes.

power element: Sensitive element of a temperature-operated control.

power factor: Correction coefficient for ac power necessary because of changing current and voltage values. The power factor of an ac circuit is a ratio of the apparent power compared to the true power. Power factor is important because utility companies charge industries large penalties for a poor power factor. The power factor of a circuit or system can be found by the equation:

$$PF = \frac{true\ power\ (W)}{apparent\ power\ (VA)}$$

power factor correction: Capacitance is used for power factor correction. When the capacitive current is at its peak positive value, the inductive current is at its peak negative value, and vice-versa. Since these two currents are in direct opposition, one can be used to cancel the other. To determine the amount of capacitance for power factor correction, first calculate the power factor (see *power factor*). Find the reactive power (vars) using the equation:

$$\text{Vars} = \sqrt{VA^2 - W^2}$$

This reactive power in the circuit can be canceled by an equal amount of capacitive vars, found by the equation $X_C = E^2/\text{vars}$. Finally, the amount of capacitance, in farads, needed to produce the correction can be found using the following equation:

$$C = \frac{1}{2 \; x \; \pi \; x \, frequency \, x \, X_C}$$

power loss (cable): Losses due to internal cable impedance, mainly I^2r; the losses cause heating.

power pool: A group of power systems operating as an interconnected system.

P-P: Peak to peak.

precast concrete: Concrete units (such as piles or vaults) cast away from the construction site and set in place.

precious metal: Gold, silver, or platinum.

premolded: A splice or termination manufactured of polymers, ready for field application.

pressure: 1) An energy impact on a unit area; force or thrust exerted on a surface. 2) Electromotive force commonly called *voltage.*

pressure motor control: A device that opens and closes an electrical circuit as pressures change.

primary: Normally referring to the part of a device or equipment connected to the power supply circuit.

primary cell: A device for transforming chemical energy into electrical energy. It consists essentially of a container with a solution (electrolyte) and two plates of electrodes.

primary coil: The coil into which the original energy is introduced and which sets up magnetic lines of force to link with another coil in which energy is induced.

primary control: Device that directly controls operation of a heating system.

printed circuit: A board having interconnecting wiring printed on its surface and designed for mounting of electronic components.

process: Path of succession of states through which a system passes.

program, computer: The ordered listing of sequence of events designed to direct the computer to accomplish a task.

propagation: The travel of waves through or along a medium.

property: An observable characteristic.

protected enclosure: Having all openings protected with screening, etc.

protector, circuit: An electrical device that will open an electrical circuit if excessive electrical conditions occur.

proton: The hydrogen atom nucleus; it is electrically positive; mass = 1.672×10^{-24} grams; charge = 0.16 attocoulomb.

prototype: The first full size working model.

proximity effect: The distortion of current density due to magnetic fields; increased by conductor diameter, close spacing, frequency, and magnetic materials such as steel conduit or beams.

PSAR (Preliminary Safety Analyses Report) (nuclear): Construction permit.

PSI: Pound force per square inch.

PT: Potential transformer.

pull box: A sheet metal box-like enclosure used in conduit runs, either single conduits or multiple conduits, to facilitate pulling in of cables from point to point in long runs, or to provide installation of conduit support bushings needed to support the weight of long riser cables, or to provide for turns in multiple-conduit runs.

pull-down: Localized reduction of conductor diameter by longitudinal stress.

pulling compound (lubricant): A substance applied to the surface of a cable to reduce the coefficient of friction during installation.

pulling eye: A device attached to a cable to facilitate field connection of pulling ropes.

pulsating current: A direct current in which the value is not constant but the flow is in one direction.

pulsating function: A periodic function whose average value over a period is not zero.

pulse: A brief excursion of a quantity from normal.

pumped storage (hydro power): The storage of power by pumping a reservoir full of water during off-peak, then depleting the water to generate when needed.

puncture: Where breakdown occurs in an insulation.

purge: To clean.

push button: 1) A device that completes an electric circuit as long as a button is depressed. 2) A switch in which electrical contacts are closed by pushing one button and are opened by pushing another.

PVC (polyvinyl chloride): A thermoplastic insulation and jacket compound.

PWR (pressurized water reactor) (nuclear power): A basic nuclear fission reaction in which water is used to transfer energy from the reactor; the water exchanges its heat with a secondary loop to produce steam for the turbine.

pyroconductivity: Electric conductivity that develops with changing temperature, and notably upon fusion, in solids that are practically non-conductive at atmospheric temperatures.

pyrometer: Thermometer that measures the radiation from a heated body.

Q

QA (Quality Assurance): All the planned and systematic actions to provide confidence that a structure, system, or component will perform satisfactorily.

QAP (nuclear): Quality assurance policy.

quadrant: The quarter of a circle, or of its circumference.

quadruplexed: Twisting of four conductors together.

qualified life (nuclear power): The period of time for which satisfactory performance can be demonstrated for a specific set of service conditions.

qualified person: A person familiar with construction, operation, and hazards.

quarter bend: A bend through an arc of 90 degrees; as in a piece of conduit. Sometimes called *elbow*. See Fig. Q-1.

quick-: Prefix indicating a device that has a high contact speed independent of the operator; example: quick-make or quick-break. See description of *quick-break switch* on page 156 for details of operation.

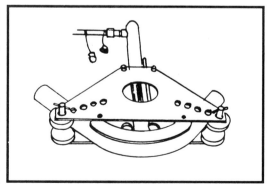

Fig. Q-1: Hydraulic pipe bender bending a 90° bend (elbow) in 4-inch rigid steel conduit.

quick-break switch: Usually of the knife-blade type. The blade is made of two pieces. As the switch is pulled out, the first half of the blade is withdrawn, and as the throw increases, the second half is drawn out by the action of a spring attached to the first blade.

quotation: A firm price given a buyer. For example, "the supplier quoted a price of $200 for a 200-amperes, three-phase disconnect switch " or "The electrical contractor's quotation for the electrical work was $17,500."

R

raceway: Any channel designed expressly for holding wire, cables, or bars and used solely for that purpose. See Fig. R-1.

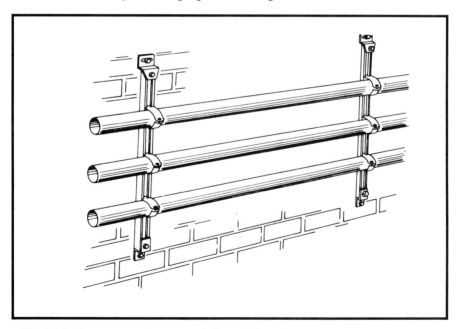

Fig. R-1: Raceway system supported on cable racks.

rack (cable): A device to support cables.

radar: A radio detecting and ranging system.

radial feeder: A feeder connected to a single source.

radian (rad): A supplementary SI unit for plane angles; the plane angle with its vertex at the center of a circle that is subtended by an arc equal in length to the radius.

radiant energy: Energy traveling in the form of electromagnetic waves.

radiant heating: Heating system in which warm or hot surfaces are used to radiate heat into the space to be conditioned.

radiation: The process of emitting radiant energy in the form of waves or particles.

radiation, blackbody: Energy given off by an ideal radiating surface at any temperature.

radiation, nuclear: The release of particles and rays during disintegration or decay of an atom's nucleus; these rays cause ionization; they are: alpha particles, beta particles, gamma rays.

radius, bending: The radii around which cables are pulled.

radius, training: The radii to which cables are bent by hand positioning, not while the cables are under tension.

rail clamp: A device to connect cable to a track rail.

rainproof: So constructed, protected, or treated as to prevent rain from interfering with the successful operation of the apparatus under specified test conditions.

rainshield: An inverted funnel to increase the creepage over a stress cone.

raintight: So constructed or protected that exposure to a beating rain will not result in the entrance of water. See Fig. R-2.

raked joint: Joint formed in brickwork by raking out some of the mortar an even distance from the face of the wall.

ram: Random access memory.

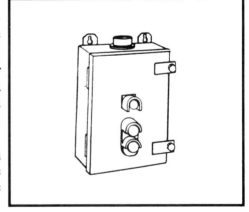

Fig. R-2: Raintight motor control.

rated: Indicating the limits of operating characteristics for application under specified conditions.

rating, temperature (cable): The highest conductor temperature attained in any part of the circuit during: a) normal operation, b) emergency overload, c) short circuit.

rating, voltage: The thickness of insulation necessary to confine voltage to a cable conductor after withstanding the rigors of cable installation and normal operating environment.

REA (Rural Electrification Administration): A federally supported program to provide electrical utilities in rural areas.

reactance: 1) The imaginary part of impedance. 2) The opposition to ac due to capacitance (X_C) and/or inductance (X_L).

reactor: A device to introduce capacitive or inductive reactance into a circuit.

reactor, nuclear: An assembly designed for a sustained nuclear chain reaction; the chain reaction occurs when the mass of the fuel reaches a critical value having enough free neutrons or enough heat to sustain fusion.

real time: The actual time during which a physical process transpires.

reamer: A finishing tool that finishes a circular hole more accurately than a drill. A tool for removing burrs on the inside of the mouth on metal conduit. See Fig. R-3.

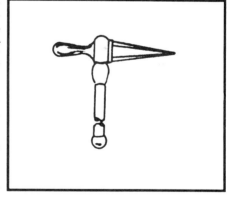

receptacle: A contact device installed at an outlet for the connection of an attachment plug and flexible cord to supply portable equipment.

reciprocating: Action in which the motion is back and forth in a straight line.

Fig. R-3: A pipe reamer is used to remove burrs from the inside of metal conduit.

recognized component: An item to be used as a subcomponent and tested for safety by an independent testing laboratory.

recorder: A device that makes a permanent record, usually visual, of varying signals.

rectifiers: Devices used to change alternating current to direct current.

rectify: To change from ac to dc.

red-leg: See *high-leg.*

redraw: Drawing of wire through consecutive dies.

reducing joint: A splice of two different size conductors.

reduction: The gain of electrons by a constituent of a chemical reaction.

redundancy: The use of auxiliary items to perform the same functions for the purpose of improving reliability and safety.

reel: A drum having flanges on the ends; reels are used for wire/cable storage.

reflective insulation: Thin sheets of metal or foil on paper set in the exterior walls of a building to reflect radiant energy.

refrigerant: Substance used in refrigerating mechanisms to absorb heat in an evaporator coil and to release heat in a condenser as the substance goes from a gaseous state back to a liquid state.

register: Combination grille and damper assembly covering on an air opening or end of an air duct.

regulation: The maximum amount that a power supply output will change as a result of the specified change in line voltage, output load, temperature, or time.

reinforced jacket: A cable jacket with reinforcing fiber between layers.

relative capacitance: The ratio of a material's capacitance to that of a vacuum of the same configuration; will vary with frequency and temperature.

relative humidity: Ratio of amount of water vapor present in air to greatest amount possible at same temperature.

relay: A device designed to abruptly change a circuit because of a specified control input. See Fig. R-4 on page 161.

relay, overcurrent: A relay designed to open a circuit when current in excess of a particular setting flows through the sensor.

reliability: The probability that a device will function without failure over a specified time period or amount of usage.

relief valve: Safety device to permit the escape of steam or hot water subjected to excessive pressures or temperatures.

remote-control circuits: The control of a circuit through relays, etc.

repeatability: The closeness of agreement among repeated measurements of the same variable under the same conditions.

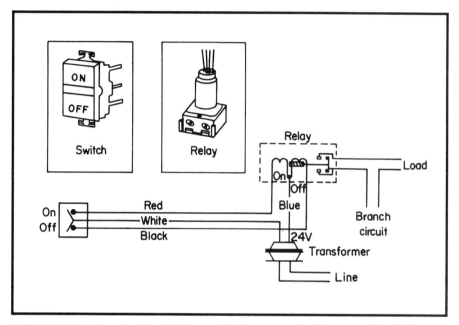

Fig. R-4: A relay circuit with the required components for proper functioning.

reproducibility: The ability of a system or element to maintain its output or input over a relatively long period of time.

reservoir, thermal: A body to which and from which heat can be transferred indefinitely without a change in the temperature of the reservoir.

residual elements: Elements present in an alloy in small quantities, but not added intentionally.

residual stress: Stress present in a body that is free of external forces or thermal gradients.

resin: The polymeric base of all jacketing, insulating, etc.; compounds, both rubber and plastic.

resistance: The opposition in a conductor to current; the real part of impedance.

resistance furnace: A furnace that heats by the flow current against ohmic resistance internal to the furnace.

resistance, thermal: The opposition to heat flow; for cables it is expressed by degrees centigrade per watt per foot of cable.

resistance welding: Welding by pressure and heat when the work piece's resistance in an electric circuit produces the heat.

resistivity: A material characteristic opposing the flow of energy through the material; expressed as a constant for each material; it is affected by temper, temperature, contamination, alloying, coating, etc.

resistor: A device whose primary purpose is to introduce resistance.

resistor, bleeder: 1) Used to drain current after a device is de-energized. 2) Improves voltage regulation. 3) Protects against voltage surges.

resolution: The degree to which nearly equal values of a quantity can be discriminated.

resolver: A device whose input and output is a vector quantity.

resonance: A condition reached in an electrical circuit when the inductive reactance neutralizes the capacitance reactance leaving ohmic resistance as the only opposition tot he flow of current.

resonating: The maximizing or minimizing of the amplitude or other characteristics provided the maximum or minimum is of interest.

response: A quantitative expression of the output as a function of the input under conditions that must be explicitly stated.

restrike: A resumption of current between contacts during an opening operation after an interval of zero current of ¼ cycle at normal frequency or longer.

reverse lay: Reversing the direction of lay about every five feet during cabling of aerial cable to facilitate field connections.

reversible process: Can be reversed and leaves no change in system or surroundings.

RF: Radio frequency: 10kGz to GHz.

RFI: Radio frequency interference.

rheology: The science of the flow and deformation of matter.

rheostat: A variable resistor that can be varied while energized, normally one used in a power circuit.

ring-out: 1) A circular section of insulation or jacket. 2) The continuity testing of a conductor.

ripple: The ac component from a dc power supply arising from sources within the power supply.

riser: A vertical run of conductors in conduit or busway, for carrying electrical power from one level to another in a building.

riser diagram: Electrical drawing that provides an instant overview of the electric service, related feeders, and subpanels. See Fig. R-5.

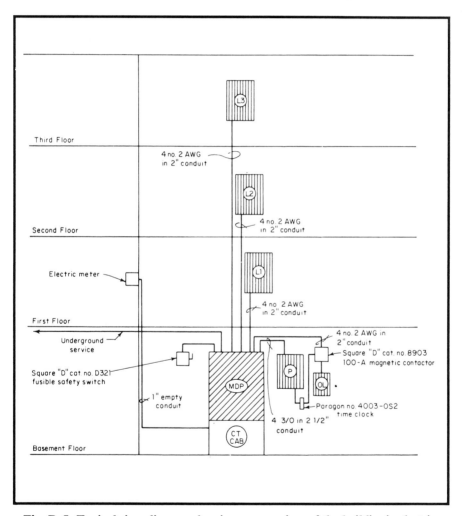

Fig. R-5: Typical riser diagram showing an overview of the building's electric service and related components.

riser valve: Device used to manually control flow of refrigerant in vertical piping.

rivet: A short, metal, bolt-like fastening, without threads, which is clinched by hammering. Rivets are designated by the shape of the beads as flat or panhead, buttonhead, countersink, mushroom, soap, or swollen neck.

RMS (Root-mean-square): The square root of the average of the square of the function taken throughout the period.

rock duster: A machine to distribute rock dust over coal to prevent dust explosions.

rod: The shape of solidified metal convenient for wire drawing, usually $\frac{5}{16}$ inch or larger.

roller bearing: A bearing made of hardened-steel rollers instead of the round steel balls used in ball bearings.

rolling: Reducing the cross-sectional area of metal stock or otherwise shaping metal products using rotating rolls.

ROM: Read only memory.

Romex: General Cable's trade name for type NM cable; but it is used generically by electrical workers to refer to any nonmetallic sheathed cable. See Fig. R-6. Also see Fig. R-7 on page 165.

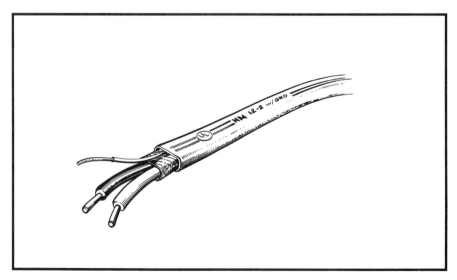

Fig. R-6: Romex or type NM cable is mostly used for residential and small commerical wiring systems.

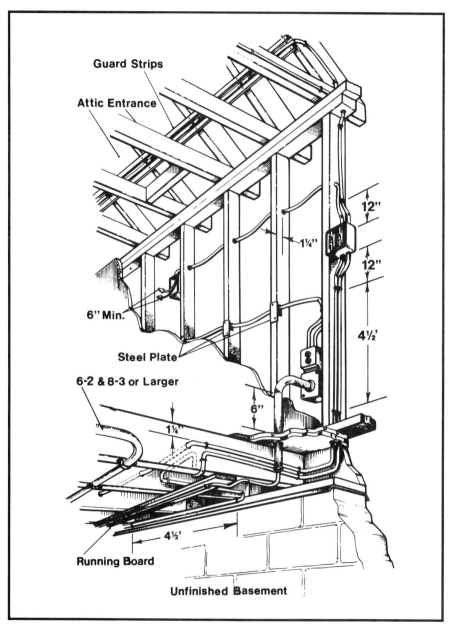

Fig. R-7: NE Code rules summarizing the installation of type NM (Romex) cable.

rope job: An installation of non-metallic sheathed cable.

roster: A tabulated schedule or list of names.

rotary: Turning on its axis, like a wheel.

rotary blower: An encased rotating fan such as that used for forced draft in furnaces.

rotary converter: A single machine connected to an ac circuit that delivers dc or vice versa.

rotor: Rotating part of a mechanism.

roughing in: The first stage of an electrical installation, when the raceway, cable, wires, boxes and other equipment are installed; that electrical work which must be done before any finishing or cover-up phases of building construction can be undertaken.

rough inspection: The first inspection made of an electrical installation after the conductors, boxes, and other equipment have been installed in a building under construction.

round off: To delete the least significant digits of a numeral and adjust the remaining by given rules.

RPM: Revolutions per minute.

rubber, chlorosulfonated polyethylene (CP): A synthetic rubber insulation and jacket compound developed by DuPont as Hypalon®.

rubber, ethylene propylene: A synthetic rubber insulation having excellent electrical properties.

running board: 1) A device to permit stringing more than one conductor simultaneously. 2) Board used as protection for Romex cable in a residential attic or basement. For example,

rupture stress: The unit stress at the time of failure.

S

sacrificial protection: Prevention of corrosion by coupling a metal to an electrochemically more active metal that is sacrificed.

SAE: Society of Automotive Engineers.

safety conductor: A safety sling used during overhead line construction.

safety control: A device that will stop the refrigerating unit if unsafe pressures and/or temperatures are reached.

safety factor: The ratio of the maximum stress that something can withstand to the estimated stress that it can withstand.

safety motor control: Electrical device used to open a circuit if the temperature, pressure, and/or the current flow exceed safe conditions.

safety plug: Device that will release the contents of a container above normal pressure conditions and before rupture pressures are reached.

safety switch: A knive-blade switch enclosed in a metal box and operated externally.

sag: The difference in elevation of a suspended conductor.

sag, apparent: Sag between two points at 60°F and no wind.

sag, final: Sag under specified conditions after the conductor has been externally loaded, then the load removed.

sag, initial: Sag prior to external loading.

sag, maximum: Sag at midpoint between two supports.

sag section: Conductor between two snubs.

sag snub: Where a conductor is held fixed and the other end moved to adjust sag.

sag, total: Sag under ice loading.

sampling: A small quantity taken as a sample for inspection or analysis.

saturation: The condition existing in a circuit when an increase in the driving signal does not produce any further change in the resultant effect. A magnetic material is said to be saturated when, upon increasing the ampere turns, no increase in the number of magnetic lines of force is obtained.

scalar: A quantity (as mass or time) that has a magnitude described by a real number and no direction.

scan: To examine sequentially, part by part.

scavenger pump: Mechanism used to remove fluid from sump or containers.

scintillation: The optical photons emitted as a result of the incidence of ionization radiation.

scope: Slang for oscilloscope.

scram (nuclear): The rapid shutdown of a nuclear reactor.

screen pack: A metal screen used for straining.

SE: Service entrance.

sealed: Preventing entrance.

sealed motor compressor: A mechanical compressor consisting of a compressor and a motor, both of which are en-closed in the same sealed housing, with no external shaft or shaft seals, and with the motor operating in the refrigerant atmosphere.

sealing compound: The material poured into an electrical fitting to seal and minimize the passage of vapors. A seal-off used in hazardous locations and filled with sealing compound is shown in Fig. S-1.

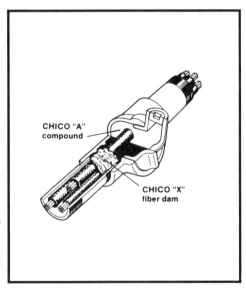

CHICO "A" compound

CHICO "X" fiber dam

Fig. S-1: Seal-off with sealing compound.

secondary: The second circuit of a device or equipment, which is not normally connected to the supply circuit. In a step-down transformer, the high-voltage side is called *primary* while the low-voltage side is called *secondary.*

Seebeck Effect: The generation of a voltage by a temperature difference between the junctions in a circuit composed of two homogeneous electrical conductors of dissimilar composition; or in a nonhomogeneous conductor the voltage produced by a temperature gradient in a nonhomogeneous region.

self-excitation: Direct current obtained from the brushes of a dc generator to provide current for its electromagnetic field. In alternators the term refers to a dc generator built on the alternator shaft. Used to provide direct current for the alternator field.

self-excited alternator: An ac generator that also produces a direct current for magnetizing its main fields.

self inductance: Magnetic field induced in the conductor carrying the current.

semiconductor: A material that has electrical properties of current flow between a conductor and an insulator.

sensible heat: Heat that causes a change in temperature of a substance.

sensor: A material or device that goes through a physical change or an electronic characteristic change as conditions change.

separable insulated connector: An insulated device to facilitate power cable connections and separations.

separately excited: A machine that gets the current needed for the excitation of its field from some outside source.

separator: Material used to maintain physical spacing between elements in cables, such as: a layer of tape to prevent jacket sticking to individual conductors.

sequence controls: Devices that act in series or in time order.

series: When two or more electrical components are connected so that the current feeding one must pass through the others, they are said to be connected in series (one after another, as in a string). See Fig. S-2 on page 170. The following four rules state the condition that exists in a series circuit.

- The current is the same in all parts of a series circuit.

- The total resistance in a series circuit is equal to the sum of the individual resistances.

- The total voltage applied to a series circuit divides between the resistors in direct proportion to their resistance.

- The sum of the voltage crops across all the resistors in a series resistive circuit is equal to the applied (source) voltage.

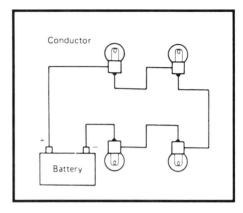

Fig. S-2: A series circuit consisting of a voltage source (battery) and four lamps connected in series by means of wire conductors.

series parallel circuit: A circuit made up of two or more simple parallel circuits all joined in series.

series-wound motor: A dc motor with the armature and field connected in series. Used on machines where variable loads occur. The speed varies with the load.

service: The equipment used to connect to the conductors run from the utility line, including metering, switching and protective devices; also the electric power delivered to the premises, rated in voltage and amperes, such as a "100-amp, 120/240-volt, three-wire service." See Fig. S-3, page 171.

service cable: The service conductors made up in the form of a cable.

service conductors: The supply conductors that extend from the street main or transformers to the service equipment of the premises being supplied.

service drop: Run of cables from the power company's aerial power lines to the point of connection to a customer's premises.

service entrance: The point at which power is supplied to a building, including the equipment used for this purpose (service main switch or panel or switchboard, metering devices, overcurrent protective devices, conductors for connecting to the power company's conductors and raceways for such conductors). See Fig. S-3, page 171.

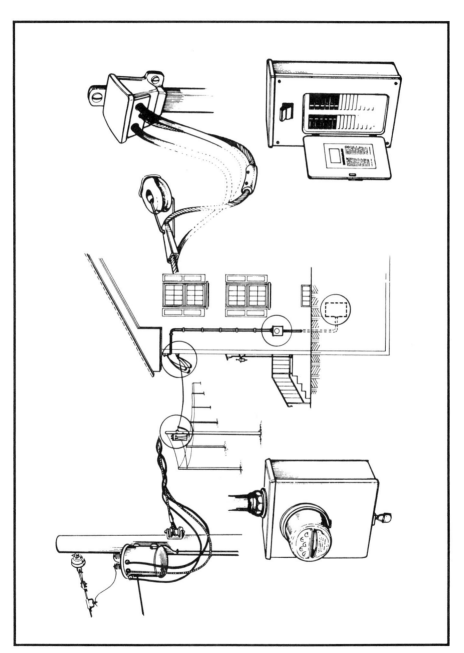

Fig. S-3: Major parts of an electric service.

171

service equipment: The necessary equipment, usually consisting of a circuit breaker or switch and fuses and their accessories, located near the point of entrance of supply conductors to a building and intended to constitute the main control and cutoff means for the supply to the building.

service lateral: The underground service conductors between the street main, including any risers at a pole or other structure or from transformers, and the first point of connection to the service-entrance conductors in a terminal box, meter, or other enclosure with adequate space, inside or outside the building wall. Where there is no terminal box, meter, or other enclosure with adequate space, the point of connection is the entrance point of the service conductors into the building.

service raceway: The rigid metal conduit, electrical metallic tubing, or other raceway that encloses the service entrance conductors. See Fig. S-4.

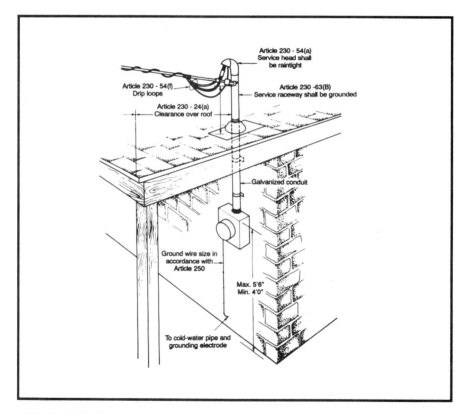

Fig. S-4: Rigid metal conduit used as service mast for a through-roof application.

service valve: A device, attached to a refrigeration system, that provides an opening for gauges and/or charging lines.

serving: A layer of helically applied material.

servomechanism: A feedback control system in which at least one of the system signals represents mechanical motion.

setting (of circuit breaker): The value of the current at which the circuit breaker is set to trip.

shaded-pole motor: A small ac motor that utilizes a shaded coil; used for light-start loads; has no brushes or commutator.

shaft furnace: A furnace used for pouring wire bars from continuous melting of cathodes.

shall: Mandatory requirement of the National Electrical Code.

shaving: Removing about 0.001 inch of metal surface.

shear: The lateral displacement in a body due to an external force causing sliding action.

sheath: A metallic close-fitting protective covering.

shield: The conducting barrier against electromagnetic fields.

shield, braid: A shield of interwoven small wires.

shield, insulation: An electrically conducting layer to provide a smooth surface in intimate contact with the insulation outer surface; used to eliminate electrostatic charges external to the shield, and to provide a fixed known path to ground.

shield, tape: The insulation shielding system whose current carrying component is thin metallic tapes, now normally used in conjunction with a conducting layer of tapes or extruded polymer.

shim: Thin piece of material used to bring members to an even or level bearing.

shore feeder: From ship to shore feeder.

shore hardness: A measure of the hardness of a plastic.

short-circuit: An often unintended low-resistance path through which current flows around, rather than through, a component or circuit.

short cycling: Refrigerating system that starts and stops more frequently than it should.

short time (nuclear power): Operation for 2 hours out of a 24-hour period.

short-time overload rating (nuclear): The limiting overload current that one third (must be at least three) of the conductors in an assembly through a penetration can carry, with all other conductors fully loaded.

shrinkable tubing: A tubing that may be reduced in size by applying heat or solvents.

shroud: Housing over a condenser or evaporator.

shunt: A device having appreciable resistance or impedance connected in parallel across other devices or apparatus to divert some of the current; appreciable voltage exists across the shunt and appreciable current may exist in it.

shunt-wound motor: Used when the motor speed must be constant, irrespective of variation in load.

sidewall load: The normal force exerted on a cable under tension at a bend; quite often called sidewall pressure.

signal: A detectable physical quantity or impulse (as a voltage, current, or magnetic field strength) by which messages or information can be transmitted.

signal circuit: Any electrical circuit supplying energy to an appliance that gives a recognizable signal. A simple residential doorbell circuit is shown in Fig. S-5.

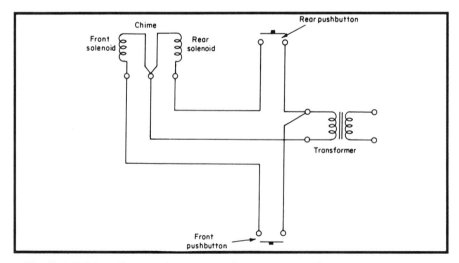

Fig. S-5: Wiring diagram showing a typical two-note chime controlled at two different locations.

silica gel: Chemical compound used as a drier.

silicon controlled rectifier (SCR): Electronic semiconductor that contains silicon.

sill: Horizontal timber forming the lowest member of a wood frame house; lowest member of a window frame.

similar poles: When two magnetic poles repel each other, they are magnetically similar or like.

sine wave, ac: Wave form of single frequency alternating current; wave whose displacement is the sine of the angle proportional to time or distance. See Fig. S-6.

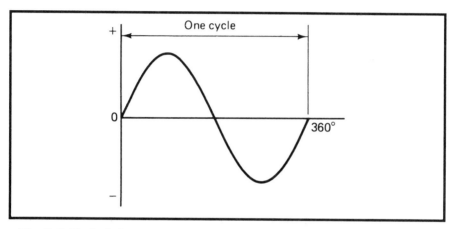

Fig. S-6: Typical sine wave.

single-phase circuit: An ac circuit having one source voltage supplied over two conductors.

single-phase motor: Electric motor that operates on single-phase alternating current.

single-phasing: The abnormal operation of a three-phase machine when its supply is changed by accidental opening of one conductor.

single-pole switch: A switch that opens and closes only one side of a circuit.

sintering: Forming articles from fusible powders by pressing the powder just under its melting point.

skin effect: The tendency of current to crowd toward the outer surface of a conductor; increases with conductor diameter and frequency.

skinning: Removing insulation from electrical conductors before making splices or connections.

slip: The difference between the speed of a rotating magnetic field and the speed of its rotor.

slippercoat: A surface lubricant factory applied to a cable to facilitate pulling, and to prevent jacket sticking.

slip rings: The means by which the current is conducted to a revolving electrical circuit.

sliver: A defect consisting of a very thin elongated piece of metal attached by only one end to the parent metal into whose surface it has been rolled.

slot: A channel opening in the stator or rotor of a rotating machine for ventilation and the insertion of windings.

soap: Slang for pulling compound.

socket: The receptacle into which the threaded portion of an incandescent lamp or plug is fitted.

soffit: Underside of a stair, arch, or cornice.

soft drawn: 1) A relative measure of the tensile strength of a conductor. 2) Wire that has been annealed to remove the effects of cold working. 3) Drawn to a low tensile.

solar cell: The direct conversion of electromagnetic radiation into electricity; certain combinations of transparent conducting films separated by thin layers of semiconducting materials.

solar heat: Heat from visible and invisible energy waves from the sun.

solder: To braze with tin alloy.

solenoid: Electric conductor wound as a helix with a small pitch; coil.

solidly grounded: No intentional impedance in the grounding circuit.

solid state: A device, circuit, or system that does not depend upon physical movement of solids, liquids, gases or plasma.

solid type PI cable: A pressure cable without constant pressure controls.

soluble oil: Specially prepared oil whose water emulsion is used as a curing grinding or drawing lubricant.

solution: 1) Homogenous mixture of two or more components. 2) The solving of a problem.

SP: Single pole.

space heater: A heater for occupied spaces.

span: A conductor between two consecutive supports.

spark: A brilliantly luminous flow of electricity of short duration that characterizes an electrical breakdown.

spark gap: Any short air space between two conductors.

spark test: A voltage withstand test on a cable while in production with the cable moving; a simple way to test long lengths of cable.

SPDT: Single-pole double-throw.

specific heat: Ratio of the quantity of heat required to raise the temperature of a body one degree to that required to raise the temperature of an equal mass of water one degree.

specs: Abbreviation for the word "specifications," which is the written precise description of the scope and details of an electrical installation and the equipment to be used in the system. A sample specification is shown in Fig. S-7 on page 178.

spectrum: The distribution of the amplitude (and sometimes phase) of the components of the wave as a function of frequency.

spike: A pulse having great magnitude.

splice: The electrical and mechanical connection between two pieces of cable. See *Western Union splice, fixture splice*, etc.

splice tube: The movable section of vulcanizing tube at the extruder.

split fitting: A conduit fitting that may be installed after the wires have been installed.

split-phase motor: Motor with two stator windings. Winding in use while starting is disconnected by a centrifugal switch after the motor attains speed, then the motor operates on the other winding.

split system: Refrigeration or air conditioning installation that places the condensing unit outside or remote from the evaporator. It is also applicable to heat pump installations.

split-wire: A way of wiring a duplex receptacle outlet with a 3-wire, 120/240 volt single-phase circuit so that one hot leg and the grounded conductor (neutral) feed one of the receptacle outlets, and the other hot leg and grounded conductor (neutral) feed the other receptacle outlet. This gives the capacity of two separate circuits to one duplex receptacle.

spray cooling: Method of refrigerating by spraying refrigerant inside the evaporator or by spraying refrigerated water.

SPST: Single-pole single-throw.

DIVISION 16-ELECTRICAL

Section 16A-General Provisions

1.The "Instruction to Bidders," "General Conditions," and "General Requirements" of the architectural specifications govern work under this Section.

2.It is understood and agreed that the Electrical Contractor has, by careful examination of the Plans and Specifications, and the site where appropriate, satisfied himself as to the nature and location of the work, and all conditions which must be met in order to carry out the work under this Section of the Contract.

3. Scope of the Work

a.The scope of the work consists of the furnishing and installing of complete electrical systems--exterior and interior--including miscellaneous systems. The Electrical Contractor shall provide all supervision, labor, materials, equipment, machinery, and any and all other items necessary to complete the systems. The Electrical Contractor shall note that all items of equipment are specified in the singular; however, the Contractor shall provide and install the number of items of equipment as indicated on the drawings and as required for complete systems.

b.It is the intention of the Specifications and Drawings to call for finished work, tested, and ready for operation.

c.Any apparatus, appliance, material, or work not shown on drawings but mentioned in the specifications or vice versa, or any incidental accessories necessary to make the work complete and perfect in all respects and ready for operation, even if not particularly specified, shall be furnished, delivered, and installed by the Contractor without additional expense to the Owner.

d.Minor details not usually shown or specified, but necessary for proper installation and operation, shall

Fig. S-7: Sample page from an electrical specification.

spurious response: Any response other than the desired response of an electric transducer or device.

squirrel-cage motor: An induction motor having the primary winding (usually the stator) connected to the power and a current is induced in the secondary cage winding (usually the rotor).

SSR: Solid state relay.

stability factor: Percent change in dissipation factor with respect to time.

stack: Any vertical line of soil, waste, or vent piping.

standard conditions: Temperature of 68 degrees Fahrenheit, pressure of 29.92 inches of mercury, and relative humidity of 30 percent.

standard deviation: 1) A measure of data from the average. 2) The root mean square of the individual deviations from the average.

standard reference position: The nonoperated or de-energized condition.

standing wave: A wave in which, for any component of the field, the ratio of its instantaneous value at one point to that at any other point does not vary with time.

standoff: An insulated support.

star connection: Three-phase generators and transformers have three coils that may be connected in star, Y, or delta. When one terminal of each coil is connected together and the other three terminals are brought out separately, the connection is called *star* or *Y*.

starter: 1) An electric controller for accelerating a motor from rest to normal speed and for stopping the motor. 2) A device used to start an electric discharge lamp. See Fig. S-8.

starting relay: An electrical device that connects and/or disconnects the starting winding of an electric motor.

starting torque: The turning effort produced by a motor upon its shaft through the electro-magnetic effect at the initial flow of current.

starting winding: Winding in an electric motor used only during the brief period when the motor is starting.

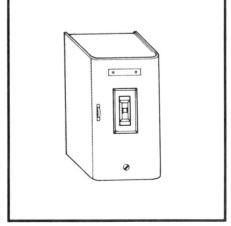

Fig. S-8: Simple motor starter.

static: Interference caused by electrical disturbances in the atmosphere.

stator: The portion of a rotating machine that includes and supports the stationary active parts.

steady state: A characteristic exhibiting only negligible change over a long period of time.

steam: Water in vapor state.

steam heating: Heating system in which steam from a boiler is conducted to radiators in a space to be heated.

steradian: The supplemental SI unit for solid angle: the three dimensional angle having its vertex at the center of a sphere and including the area of the spherical surface equal to that of a square with sides equal in length to the radius of the sphere.

storage cell: One of the sections of a storage battery.

strain: 1) A change in characteristic resulting from external forces. 2) To screen foreign materials from a substance.

strand: A group of wires, usually twisted or braided.

strand, annular: A concentric conductor over a core; used for large conductors (1000 kcmil @ 60 Hertz) to make use of skin effect; core may be of rope, or twisted I-beam.

strand, bunch: A substrand for a rope-lay conductor; the wires in the substrand are stranded simultaneously with the same direction; bunched conductors flex easily and with little stress.

strand, class: A system to indicate the type of stranding; the postscripts are alpha.

strand, combination: A concentric strand having the outlet layer of different size; done to provide smoother outer surface; wires are sized with +5% tolerance from nominal.

strand, compact: A concentric stranding made to a specified diameter of 8%-10% less than standard by using smaller than normal closing die, and for larger sizes, preshaping the strands for the outer layer(s).

strand, compressed: The making of a tight stranded conductor by using a small closing die.

strand, concentric: Having a core surrounded by one or more layers of helically laid wires each of one size, each layer increased by six.

strand, herringbone lay: Adjacent bunches having opposite direction of lay in a layer of a rope-lay cable.

strand, nonspecular: One having a treated surface to reduce light reflection.

strand, regular lay: Rope stranding having left-hand lay within the substrands and right-hand lay for the conductor.

strand, reverse-lay: A stranding having alternate direction of lay for each layer.

strand, rope-lay: A conductor having a lay-up of substrands; substrand groups are bunched or concentric.

strand, sector: A stranded conductor formed into sectors of a circle to reduce the overall diameter of a cable.

strand, segmental: One having sectors of the stranded conductor formed and insulated one from the other, operated in parallel; used to reduce ac resistance in single conductor cables.

strand, unilay (unidirectional): Stranding having the same direction of lay for all layers; used to reduce diameter, but is more prone to birdcaging.

stratification of air: Condition in which there is little or no air movement in the room; air lies in temperature layers.

stress: 1) An internal force set up within a body to resist or hold it in equilibrium. 2) The externally applied forces.

stress-relief cone: Mechanical element to relieve the electrical stress at a shield cable termination; used above 2kV.

striking: The process of establishing an arc or a spark.

striking distance: The effective distance between two conductors separated by an insulating fluid such as air.

stringers: Members supporting the treads and risers of a stair.

strip: To remove insulation or jacket.

strip cooler: A device to cool strips of compound.

strut: A compression member other than a column or pedestal.

studs: Vertically set skeleton members of a partition or wall to which the lath is nailed.

sub-panel: A panelboard in a residential system that is fed from the service panel; or any panel in any system that is fed from another, or main, panel supplied by a circuit from another panel.

substation: An assembly of devices and apparatus to monitor, control, transform, or modify electrical power. See Fig. S-9 on page 182.

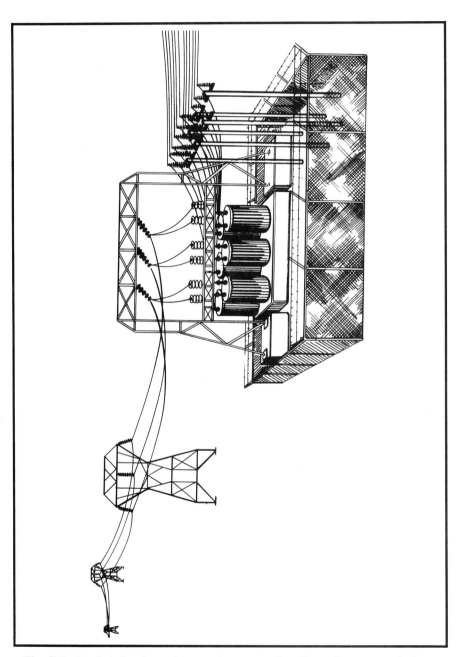

Fig. S-9: A substation is used to monitor, control, and transform transmission power.

superconductors: Materials whose resistance and magnetic permeability are infinitesimal at absolute zero (273°C).

supervised circuit: A closed circuit having a current responsive device to indicate a break or ground.

surge: 1) A sudden increase in voltage and current. 2) Transient condition.

surge arrestor: A device used on electrical circuits to detect and check sudden increases in voltage and current.

switch: A device for opening and closing or for changing the connection of a circuit.

switch, ac general-use snap: A general-use snap switch suitable only for use on alternating current circuits and for controlling the following:

- Resistive and inductive loads (including electric discharge lamps) not exceeding the ampere rating at the voltage involved.

- Tungsten-filament lamp loads not exceeding the ampere rating of the switches at the rated voltage.

- Motor loads not exceeding 80 percent of the ampere rating of the switches at the rated voltage.

switch, ac-dc general-use snap: A type of general-use snap switch suitable for use on either direct or alternating current circuits and for controlling the following:

- Resistive loads not exceeding the ampere rating at the voltage involved.

- Inductive loads not exceeding one-half the ampere rating at the voltage involved, except that switches having a marked horsepower rating are suitable for controlling motors not exceeding the horsepower rating of the switch at the voltage involved.

- Tungsten-filament lamp loads not exceeding the ampere rating at 125 volts, when marked with the letter T.

switchboard: A large single panel, frame, or assembly of panels having switches, overcurrent, and other protective devices, buses, and usually instruments mounted on the face or back or both. Switchboards are generally accessible from the rear and from the front and are not intended to be installed in cabinets.

switch, general-use: A switch intended for use in general distribution and branch circuits. It is rated in amperes and is capable of interrupting its rated voltage.

switch, general-use snap: A type of general-use switch so constructed that it can be installed in flush device boxes or on outlet covers, or otherwise used in conjunction with wiring systems recognized by the National Electrical Code.

switch, isolating: A switch intended for isolating an electrical circuit from the source of power. It has no interrupting rating and is intended to be operated only after the circuit has been opened by some other means.

switch, knife: A switch in which the circuit is closed by a moving blade engaging contact clips.

switch-leg: That part of a circuit run from a lighting outlet box where a luminaire or lampholder is installed down to an outlet box that contains the wall switch that turns the light or other load on or off; it is a control leg of the branch circuit.

switch, motor-circuit: A switch, rated in horsepower, capable of interrupting the maximum operating overload current of a motor having the same horsepower rating as the switch at the rated voltage. See Fig. S-10 on page 185.

symmetrical: Exhibiting symmetry.

symmetry: The correspondence in size, form, and arrangement of parts on opposite sides of a plane or line or point.

synchronism: Connected ac systems, machines, or a combination operating at the same frequency when the phase angle displacements between voltages in them are constant, or vary about a steady and stable average value.

synchronous: Simultaneous in action and in time (in phase).

synchronous machine: A machine in which the average speed of normal operation is exactly proportional to the frequency of the system to which it is connected.

synchronous motor: A motor that maintains a constant speed as long as the speed of the generator supplying it remains constant.

ENCLOSURE	EXPLANATION
NEMA Type 1 General Purpose	To prevent accidental contact with enclosed apparatus. Suitable for application indoors where not exposed to unusual service conditions.
NEMA Type 3 Weatherproof (Weather Resistant)	Protection against specified weather hazards. Suitable for use outdoors.
NEMA Type 3R Raintight	Protects against entrance of water from a rain. Suitable for general outdoor application not requiring sleetproof.
NEMA Type 4 Watertight	Designed to exclude water applied in form of hose stream. To protect against stream of water during cleaning operations, etc.
NEMA Type 5 Dusttight	Constructed so that dust will not enter the enclosed case. Being replaced in some equipment by Square D NEMA 12 Types.
NEMA Type 7 Hazardous Locations A, B, C or D Class I — Air Break Letter or letters following type number indicates particular groups of hazardous locations per N.E.C.	Designed to meet application requirements of National Electrical Code for Class I, Hazardous locations (Explosive atmospheres). Circuit interruption occurs in air.
NEMA Type 9 Hazardous Locations E, F or G Class II Letter or letters following type number indicates particular groups of hazardous locations per N.E.C.	Designed to meet application requirements of National Electrical Code for Class II Hazardous Locations (Combustible Dusts, etc.).
NEMA Type 12 Industrial Use	For use in those industries where it is desired to exclude dust, lint, fibers and filings, or oil or coolant seepage.

Fig. S-10: Practical application of safety switches used for motors and other power circuits.

synchronous speed: The speed of rotation of the magnetic flux produced by linking the primary winding.

synchrotron: A device for accelerating charged particles to high energies in a vacuum; the particles are guided by a changing magnetic field while they are accelerated in a closed path.

system: A region of space or quantity of matter undergoing study.

T

tabulate: To arrange items or data in a table or list.

tachometer: An instrument for measuring revolutions per minute.

tackle: The chain, rope, pulleys, or blocks used for hoisting purposes in the erection of heavy work.

take-off: The procedure by which a listing is made of the numbers and types of electrical components and devices for an installation, taken from the electrical plans, drawings, and specs for the job. See take-off sheet in Fig. T-1 on page 188.

take-up: 1) A device to pull wire or cable. 2) The process of accumulating wire or cable.

tandem extrusion: Extruding two materials, the second being applied over the first, with the two extruders being just a few inches or feet apart in the process.

tangent (geometry): A line that touches a curve at a point so that it is closer to the curve in the vicinity of the point than any other line drawn through the point; (trigonometry) in a right triangle it is the ratio of the opposite to the adjacent sides for a given angle.

tank test: The immersion of a cable in water while making electrical tests; the water is used as a conducting element surrounding the cable.

tap: A splice connection of a wire to another wire (such as a feeder conductor in an auxiliary gutter) where the smaller conductor runs a short distance (usually only a few feet, but could be several feet) to supply a panelboard or motor controller or switch. Also called a "tap-off," indicating that energy is being taken from one circuit or piece of equipment to supply another circuit or load; a tool that cuts or machines threads in the side of a round hole. See Fig. T-2 on page 189.

PRICING SHEET

JOB _____ ESTIMATE NO. _____

WORK _____ SHEET NO. ____ OF ____ SHEETS

ESTIMATED BY _____ PRICED BY _____ EXTENDED BY _____ CHECKED BY_____ DATE _____

MATERIAL	MATERIAL					LABOR		
	QUANTITY	LIST PRICE	PER	DISC.	EXTENSION	UNIT	PER	EXTENSION
1								
2								
3								
4								
5								
6								
7								
8								
9								
10								
11								
12								
13								
14								
15								
16								
17								
18								
19								
20								
21								
22								
23								
24								
25								
26								
27								
28								
29								
30								
31								
32								
33								
34								
35								

FORM E-12

Fig. T-1: Take-off sheet used to list materials when estimating the cost of an electrical installation.

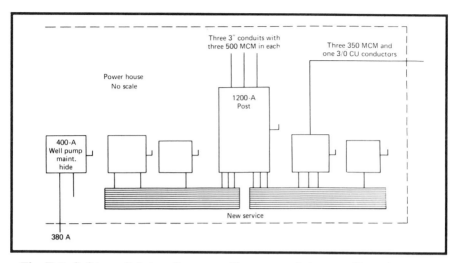

Fig. T-2: Safety-switch taps from a auxiliary gutter (wire trough).

tap drill: Drill used to form hole prior to placing threads in hole. The drill is the size of the root diameter of tap threads. See Fig. T-3.

tape: A relatively narrow, long, thin, flexible fabric, film or mat or combination thereof; helically applied tapes are used for cable insulation, especially at splices; for the first century the primary insulation for cables above 2kV was oil saturated paper tapes.

taper: A gradual and uniform decrease in size, as a tapered socket, a tapered shaft, a tapered shank.

target, sag: A visual reference used when sagging.

TC: 1) Thermocouple. 2) Time constant. 3) Timed closing.

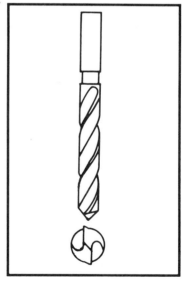

Fig. T-3: Typical tap drill.

TDR: 1) Time delay relay 2) Time domain reflectometer, pulse-echo testing of cables; signal travels through cable until impedance discontinuity is encountered, then part of signal is reflected back; distance to fault can be estimated. Useful for finding faults, broken shields, or conductor.

Technical Appeal Board (TAB) of UL: A group to recommend solutions to technical differences between UL and a UL client.

Teflon®: A DuPont trade name for polytetrafluoroethylene which is used as high temperature insulation and has low dissipation factor and low relative capacitance.

telegraphy: Telecommunication by the use of a signal code.

telemetering: Measurement with the aid of intermediate means that permits interpretation at a distance from the primary detector.

telephone: The transmission and reception of sound by electronics.

temper: A measure of the tensile strength of a conductor; indicative of the amount of annealing or cold working done to the conductor.

temperature, ambient: The temperature of the surrounding medium, such as air around a cable.

temperature, coefficient of resistance: The unit change in resistance per degree temperature change.

temperature, emergency: The temperature to which a cable can be operated for a short length of time, with some loss of useful life.

temperature humidity index: Actual temperature and humidity of a sample of air compared to air at standard conditions.

temperature, operating: The temperature at which a device is designed or rated for normal operating conditions; for cables: the maximum temperature for the conductor during normal operation.

temporary: This single word is used to mean either "temporary service" (which a power company will give to provide electric power in a building under construction) or "temporary inspection" (the inspection that a code-enforcing agency will make of a temporary service prior to inspection of the electrical work in a building under construction.) See Fig. T-4 on page 191.

tensile strength: The greatest longitudinal stress a material — such as a conductor—can withstand before rupture or failure while in service.

tension, final unloaded conductor: The tension after the conductor has been stretched for an appreciable time by loads simulating ice and wind.

tension, initial conductor: The tension prior to any external load.

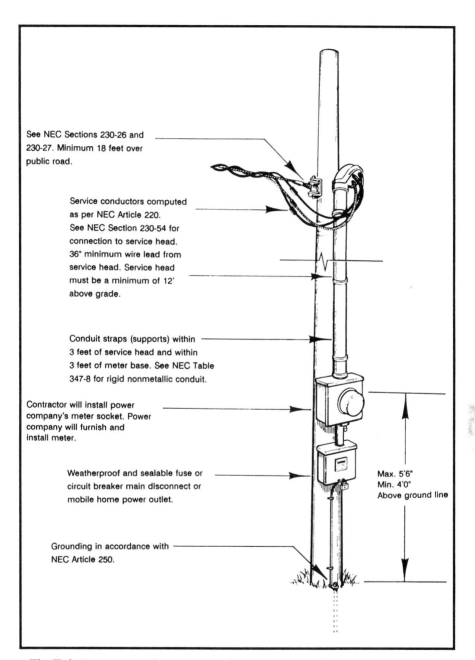

See NEC Sections 230-26 and 230-27. Minimum 18 feet over public road.

Service conductors computed as per NEC Article 220. See NEC Section 230-54 for connection to service head. 36" minimum wire lead from service head. Service head must be a minimum of 12' above grade.

Conduit straps (supports) within 3 feet of service head and within 3 feet of meter base. See NEC Table 347-8 for rigid nonmetallic conduit.

Contractor will install power company's meter socket. Power company will furnish and install meter.

Weatherproof and sealable fuse or circuit breaker main disconnect or mobile home power outlet.

Grounding in accordance with NEC Article 250.

Max. 5'6"
Min. 4'0"
Above ground line

Fig. T-4: Components for one type of temporary electric service.

tension, working: The tension that should be used for a portable cable on a power reel; it should not exceed 10% of the cable breaking strength.

terminal: A device used for connecting cables. See Fig. T-5.

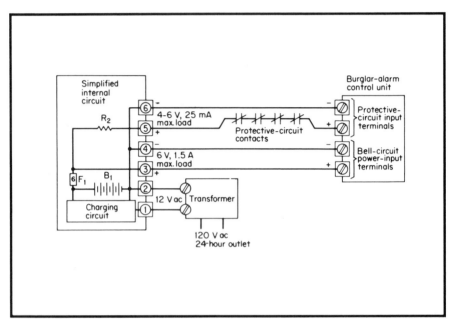

Fig. T-5: Several types of cable terminals.

termination: 1) The connection of a cable. 2) The preparation of shielded cable for connection.

testlight: Light provided with test leads that is used to test or probe electrical circuits to determine if they are energized.

test, proof: Made to demonstrate that the item is in satisfactory condition.

test, voltage breakdown: 1) Step method — applying a multiple of rated voltage to a cable for several minutes, then increasing the applied voltage by 20% for the same period until breakdown. 2) Applying a voltage at a specified rate until breakdown.

test, voltage life: Applying a multiple of rated voltages over a long time period until breakdown; time to failure is the parameter measured.

test, volume resistivity: Measuring the resistance of a material such as the conducting jacket or conductor stress control.

test, water absorption: Determination of how much water a given volume of material will absorb in a given time period; this test is being superseded by the EMA test.

therm: Quantity of heat equivalent to 100,000 Btu.

thermal cutout: An overcurrent protective device containing a heater element in addition to, and affecting, a renewable fusible member that opens the circuit. It is not designed to interrupt short-circuit currents. See Fig. T-6.

thermal endurance: The relationship between temperature and time of degrading insulation until failure, under specified conditions.

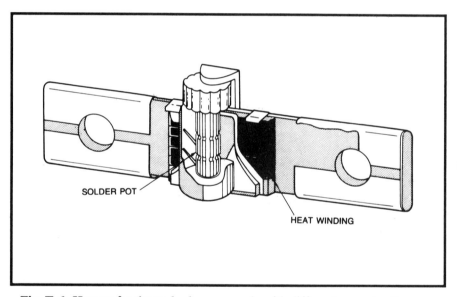

SOLDER POT

HEAT WINDING

Fig. T-6: Heaters for thermal relays are made with different current ratings, so that within its limits any starter can be used with different size motors and still provide proper protection by selecting the size of heater that corresponds to the full-load current of the motor being used.

thermally protected (as applied to motors): When the words thermally protected appear on the nameplate of a motor or motor-compressor, it means that the motor is provided with a thermal protector designed to protect the motor from overloads.

thermal protector (as applied to motors): A protective device that is assembled as an integral part of a motor or motor-compressor and that, when properly applied, protects the motor against dangerous overheating due to overload and failure to start.

thermal relay (hot wire relay): Electrical control used to actuate a refrigeration system. This system uses a wire to convert electrical energy into heat energy.

thermal shock: Subjecting something to a rapid, large temperature change.

thermionic emission: The liberation of electrons or ions from a solid or liquid as a result of its thermal energy.

thermistor: An electronic device that makes use of the change of resistivity of a semiconductor with change in temperature.

thermocouple: A device using the Seebeck effect to measure temperature.

thermodisk defrost control: Electrical switch with bimetal disk that is controlled by electrical energy.

thermodynamics: Science that deals with the relationships between heat and mechanical energy and their interconversion.

thermoelectric generator: A device interaction of a heat flow and the charge carriers in an electric circuit, and that requires, for this process, the existence of a temperature difference in the electric circuit.

thermoelectric heat pump: A device that transfers thermal energy from one body to another by the direct interaction of an electrical current and the heat flow.

thermometer: Device for measuring temperatures.

thermoplastic: Materials which, when reheated, will become pliable with no change of physical properties.

thermoset: Materials that may be molded, but when cured, undergo an irreversible chemical and physical property change.

thermostat: Device responsive to ambient temperature conditions.

thermostatic expansion valve: A control valve operated by temperature and pressure within an evaporator coil, which controls the flow of refrigerant.

thermostatic valve: Valve controlled by thermostatic elements.

thickness gauge: It is shaped somewhat like a pocketknife, and has blades varying in thickness by thousandths of an inch. It is used in adjusting parts with a desired amount of clearance.

threading: The cutting of screw threads, either internal or external.

three phase circuit: A polyphase circuit of three interrelated voltages for which the phase difference is 120°; the common form of generated power.

three-way switch: A switch that is used to control a light, or set of lights, from two different points. See Fig. T-7.

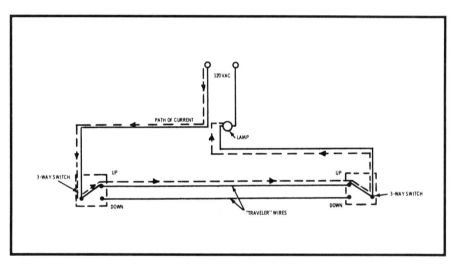

Fig. T-7: Operating characteristics of three-way switches. Operating either switch will change the current path, allowing the lamp to be controlled (turned ON or OFF) from either location.

thumper: A device used to locate faults in a cable by the release of power surges from a capacitor, characterized by the audible noise when the cable breaks down.

thyratron: A gas-filled triode tube that is used in electronic control circuits.

timer: Mechanism used to control on and off times of an electrical circuit.

timer-thermostat: Thermostat control that includes a clock mechanism. Unit automatically controls room temperature and changes it according to the time of day.

tinned: Having a thin coating of pure tin or tin alloy; the coating may keep rubber from sticking or be used to enhance connection; coatings increase the resistance of the conductor, and may contribute to corrosion by electrolysis.

title block: The outlined space usually in the lower right corner, or in strip form across the bottom of a drawing, containing name of company, title of drawing, scale, date, and other pertinent information.

toggle: A device having two stable states: i.e., toggle switch that is used to turn a circuit ON or OFF. A device may also be toggled from slow to fast, etc.

toggle bolt: Used for attaching articles to a hollow-tile wall. It consists of a screw with a swivel piece attached near the end. This piece may be swung into a length-wise position while the bolt is being inserted, after which it swings to a right-angle position, thus permitting "pulled up" on the bolt.

tolerance: The permissible variation from rated or assigned value.

topping-off: The finishing touches put to an electrical installation; mounting plates on wall switches, receptacles, and other wiring devices; installing fixtures, etc.

toroid: A coil wound in the form of a doughnut; i.e., current transformers.

torque: The turning effort of a motor.

torquing: Applying a rotating force and measuring or limiting its value.

TPE: Thermal plastic elastomer.

TPR: Thermal plastic rubber.

tracer: A means of identifying cable.

trade union: An alliance of workers organized for the purpose of securing standardized privileges for all its members.

transducer: A device by means of which energy can flow from one or more media to another.

transfer switch: A device for transferring one or more load conductor connections from one power source to another.

transformer: A static device consisting of winding(s) with or without a tap(s) with or without a magnetic core for introducing mutual coupling by induction between circuits.

transformer, potential: Designed for use in measuring high voltage; normally the secondary voltage is 120V.

transformer, power: Designed to transfer electrical power from the primary circuit to the secondary circuit(s) to 1) step-up the secondary voltage at less current or 2) step-down the secondary voltage at more current; with the voltage-current product being constant for either primary or secondary.

transformer-rectifier: Combination transformer and rectifier in which input in ac may be varied and then rectified into dc.

transformer, safety isolation: Inserted to provide a nongrounded power supply such that a grounded person accidentally coming in contact with the secondary circuit will not be electrocuted.

transformer, vault-type: Suitable for occasional submerged operation in water.

transient: 1) Lasting only a short time; existing briefly; temporary. 2) A temporary component of current existing in a circuit during adjustment to a changed load, different source voltage, or line impulse.

transistor: An active semiconductor device usually with three or more terminals. See Fig. T-8.

transmission: Transfer of electric energy from one location to another through conductors or by radiation or induction fields.

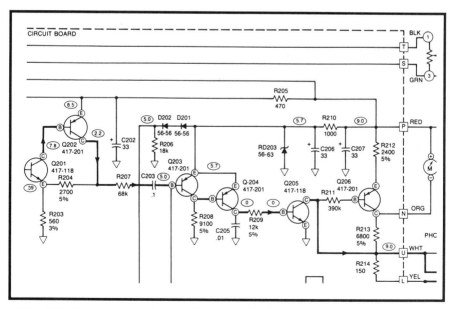

Fig. T-8: Transistors used in an electronic circuit.

transmission line: A long electrical circuit.

transposition: Interchanging position of conductors to neutralize interference.

traveler: 1)A pulley complete with suspension armor frame to be attached to overhead line structures during stringing. 2) *Travelers* are the two conductors between two three-way switches. See Fig. T-7 on page 195.

trickle charge: A rectifier changing ac to dc, and delivering same to a storage battery for a certain period of time, usually at a very minute rate.

trim: The front cover assembly for a panel, covering all live terminals and the wires in the gutters, but providing openings for the fuse cutouts or circuit breakers mounted in the panel; may include the door for the panel and also a lock.

triode: A three-electrode electron tube containing an anode, a cathode, and a control electrode.

triplex: Three cables twisted together.

trolley wire: Solid conductor designed to resist wear due to rolling or sliding current pickup trolleys.

trough: Another name for an "auxiliary gutter," which is a sheet metal enclosure of rectangular cross-section, used to supplement wiring spaces at meter centers, distribution centers, switchboards, and similar points in wiring systems where splices or taps are made to circuit conductors. The single word "gutter" is also used to refer to this type of enclosure.

trunk feeder: A feeder connecting two generating stations or a generating station and an important substation.

trussed: Framed structural pieces consisting of triangles in a single plane for supporting loads over spans.

tub: An expression sometimes used to refer to large panelboards or control center cabinets, in particular the box-like enclosure without the front trim.

tube: A hollow, long product having uniform wall thickness and uniform cross-section.

tungsten lamp: A type of incandescent lamp having a filament of fine tungsten wire. See Fig. T-9 on page 199.

tuning: The adjustment of a circuit or system to secure optimum performance.

turn: The basic coil element that forms a single conducting loop comprised of one insulated conductor.

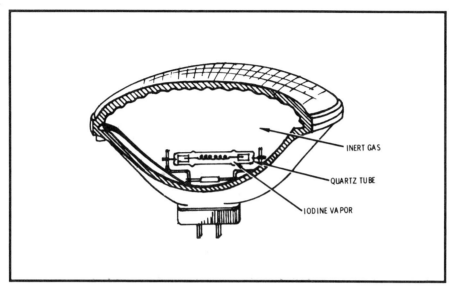

Fig. T-9: The quartz-iodine tungsten-filament lamp is basically an incandescent lamp, since ight is produced from the incandescence of its coiled tungsten filament. However, the lamp envelope, made of quartz, is filled with an iodine vapor which prevents the evaporation of the tungsten filament.

turn ratio: The ratio between the number of turns between windings in a transformer; normally primary to secondary, except for current transformers it is the ratio of the secondary to primary. See Fig. T-10.

twisted cable: A cable consisting of two or more conductors twisted together; often referred to as bell or doorbell wire. Those recently manufactured normally have a nonmetallic jacket covering the conductors. A ground or bonding wire is sometimes integrated within the jacket. Currently used for low-voltage control wiring and similar applications.

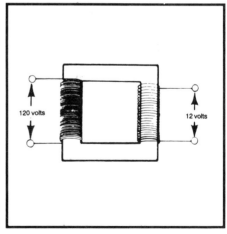

Fig. T-10: Since there are 1000 turns in the primary of this transformer and only 100 turns in the secondary, the turn ratio is 10:1.

twist test: A test to grade round material for processibility into conductors.

two-phase: A polyphase ac circuit having two interrelated voltages.

U

U bolt: A bolt shaped like the letter U threaded at both ends; used mainly to secure conduit in place. See Fig. U-1.

UHF (ultra high frequency): 300 MHz to 3GHz.

UL (Underwriters Laboratories): A nationally known laboratory for testing a product's performance with safety to the user being prime consideration; UL is an independent organization, not controlled by any manufacturer; the best known Lab for testing electrical products.

Fig. U-1: U bolt.

ultimate strength: The highest unit stress that can be sustained, this occurring just at or just before rupture.

ultrasonic: Sounds having frequencies higher than 20 KHz, which is at the upper limit of human hearing.

ultrasonic cleaning: Immersion cleaning aided by ultrasonic waves which cause microagitation.

ultrasonic detector: A device that detects ultrasonic noise such as that produced by corona or leaking gas.

ultraviolet: Radiant energy within the wave length range 10 to 380 nanometers. This energy is invisible, filtered by glass, causes teeth to glow, and causes suntan.

undercurrent: Less than normal operating current.

underground cable: A single or multiple conductor cable sheathed in lead or other waterproof materials for installation below grade. Underground cable used to feed a building electric service is shown in Fig. U-2 on page 202.

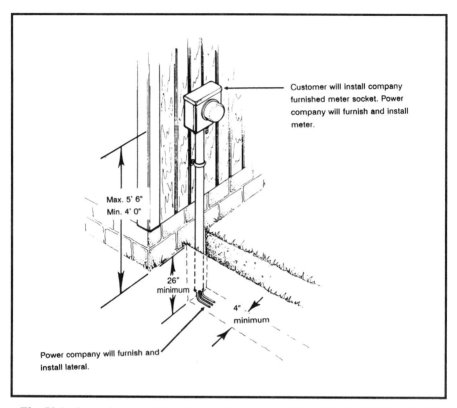

Customer will install company furnished meter socket. Power company will furnish and install meter.

Max. 5' 6"
Min. 4' 0"

26" minimum

4" minimum

Power company will furnish and install lateral.

Fig. U-2: An underground electric service (service lateral) fed with direct-burial underground conductors.

undervoltage: Less than normal operating voltage.

underwriter: Name given to representative of an inspection organization who examines electrical installations for life and fire hazards.

ungrounded: Not connected to the earth intentionally.

union: A coupling or connection for threaded conduit.

unit magnetic pole: One that will repel an equal and like pole with a force of one dyne at a distance of 1 cm. A dyne is the force which, acting upon a mass of 1 gram during 1 second, gives the mass a velocity of 1 cm per second.

unit of magnetic flux: The total number of lines of force set up in a magnetic substance. Treated as a magnetic current flowing in a magnetic circuit.

unit of magnetic intensity: Magnetomotive force. Magnetic pressure that drives lines of force through a magnetic circuit.

universal joint: A type of coupling that permits the free rotation of two shafts whose axes are not in a straight line. See Fig. U-3.

Universal motor: A motor designed to operate on either ac or dc at about the same speed and output with either. See Fig. U-4.

Urethane foam: Type of insulation that is foamed in between inner and outer walls.

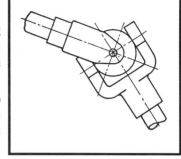

Fig. U-3: Universal joint.

URD (Underground Residential Distribution): A single phase cable usually consisting of an insulated conductor having a bare concentric neutral.

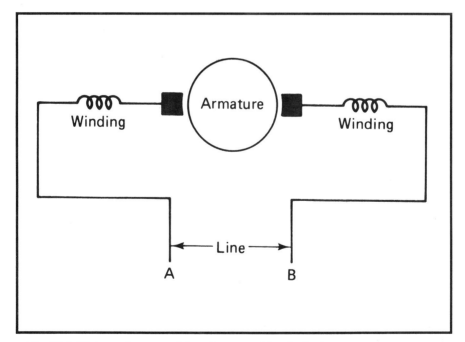

Fig. U-4: Universal motor wiring diagram with principal parts.

utility: A public service such as a telephone or electric company. A power company furnishing electric service to users is a utility.

utilization equipment: Equipment that uses electric energy for mechanical, chemical, heating, lighting, or other useful purposes. Examples would be electric motors, large and small appliances, electric heaters, air-conditioning equipment, and the like.

V

va: Volts times amps, which equals watts.

vacuum: Reduction in pressure below atmospheric pressure.

vacuum pump: Special high efficiency compressor used for creating high vacuums for testing or drying purposes.

vacuum switch: A switch with contacts in an evacuated enclosure.

vacuum tube: A device, usually in the form of a tube of glass or metal, from which the air has been exhausted and is provided with two or more elements. This sealed glass enclosure has two or more electrodes between which conduction of electricity can take place.

valley: 1) The gutter or angle formed by the meeting of two roof slopes as shown in Fig. V-1. 2) The space between vault ridges viewed from above.

valve: Device used for controlling fluid flow.

valve, expansion: Type of refrigerant control that maintains a pressure difference between high side and low side pressure in a refrigerating mechanism. The valve operates by pressure in the low or suction side.

valve, solenoid: Valve actuated by magnetic action by means of an electrically energized coil.

vapor: Word usually used to denote vaporized refrigerant rather than gas.

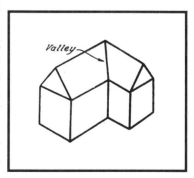

Fig. V-1: A valley is formed when two roof slopes meet.

vapor barrier: Thin plastic or metal foil sheet used in air conditioned structures to prevent water vapor from penetrating insulating material.

vaporize: To convert into a state of gas or vapor.

vapor lock: Condition where liquid is trapped in line because of a bend or improper installation that prevents the vapor from flowing.

vapor-safe: Constructed so that a device or apparatus may be operated without hazard to its surroundings in hazardous areas in aircraft.

vapor, saturated: A vapor condition that will result in condensation into liquid droplets as vapor temperature is reduced.

vapor-tight: So enclosed that vapor will not enter.

variable: A quantity that may change in value while the others remain constant.

variable resistance: A resistance that can be adjusted or changed to different values.

variable speed drive: A motor having an integral coupling device that permits the output speed of the unit to be easily varied through a continuous range.

varnish: An insulation that is applied as a liquid and is quite thin.

varying duty: That demand which requires operation of loads of varying quantity for intervals of time, both of which may be subject to wide variation.

V Block: V-shaped groove in a metal block used to hold a shaft. See Fig. V-2 on page 207.

VD: Voltage drop.

vector: A mathematical term expressing magnitude and direction.

ventilation: Circulation of air; system or means of providing fresh air.

vermiculite: Lightweight inert material resulting from expansion of mica granules at high temperatures; used as an aggregate in plaster.

vernier: An auxiliary scale permitting measurements more precise than the main scale.

vertical: Plumb, perpendicular, upright.

vibrating bell: An electric device having a clapper or hammer that strikes a bell rapidly when an electric current flows through it. It operates on the principle of electro-magnetic attraction.

vibrator coil: An induction coil so constructed that the magnetism of the core operates the make and break or vibrator of the primary circuit.

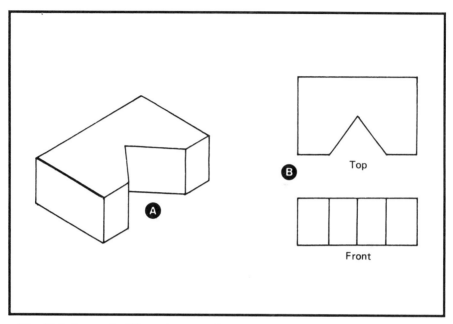

Fig. V-2: Isometric (A), and top and front (B) views of a V block, normally used in airs to hold round objects.

viscosity: Internal friction or resistance to flow of a liquid; the constant ratio of shearing stress to rate of shear.

vise: A mechanical device for holding a piece of wood or metal while it is being worked on. It consists essentially of two jaws, one fixed and one movable, the movable jaw being operated by a screw by means of which clamping action is secured. Some vises are equipped with pipe jaws for securing pipe or conduit. See Fig. V-3.

volatile flammable liquid: A flammable liquid having a flash point below 38°C or whose temperature is above its flash point.

volt: The derived SI unit for voltage: one volt equals one watt per ampere.

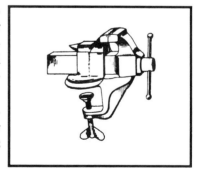

Fig. V-3: Typical bench vise.

voltage: The electrical property that provides the energy for current flow; the ratio of the work done to the value of the charge moved when a charge is moved between two points against electrical forces.

voltage, breakdown: The minimum voltage required to break down an insulation's resistance, allowing a current flow through the insulation, normally at a point.

voltage, contact: A small voltage established whenever two conductors of different materials are brought into contact; due to the difference in work functions or the ease with which electrons can cross the surface boundary in the two directions.

voltage divider: A network consisting of impedance elements connected in series to which a voltage is applied and from which one or more voltages can be obtained across any portion of the network.

voltage drop: The loss in voltage between the input to a device and the output from a device due to the internal impedance or resistance of the device. In all electrical systems, the conductors should be sized so that the voltage drop never exceeds 3 percent for power, heating, and lighting loads or combinations of these. Furthermore, the maximum total voltage drop for conductors for feeders and branch circuits combined should never exceed 5 percent.

The voltage drop in any two-wire, single-phase circuit consisting of mostly resistance-type loads, with the inductance negligible, may be found by the following equation:

$$VD + \frac{2K \times L \times I}{CM}$$

where

VD	=	drop in circuit voltage
L	=	length of conductor
I	=	current in the circuit
CM	=	area of conductor in circular mils
K	=	resistivity of conductor metal, that is, 11 for copper and 18 for aluminum

voltage, EHV (extra high voltage): 230 - 765 kV.

voltage, induced: A voltage produced in a conductor by a change in magnetic flux linking that path.

voltage rating of a cable: Phase-to-phase ac voltage when energized by a balanced three phase circuit having a solidly grounded neutral.

voltage regulator: A device to decrease voltage fluctuations to loads.

voltage, signal: Voltages to 50 V.

voltage to ground: The voltage between an energized conductor and earth.

voltage, UHV (ultra-high voltage): 765 + kV.

voltanic cell: Primary cell. Name given to the cell first discovered by Volta. Sometimes called a galvanic cell. It is a cell in which two dissimilar metals are immersed in a solution that is capable of acting chemically more upon one than on the other, to produce a difference of potential (voltage) across the metals.

voltmeter: An instrument for measuring voltage. See Fig. V-4.

VOM (volt-ohm-multimeter): A commonly used electrical test instrument to test voltage, current, resistance, and continuity.

vortex tube: Mechanism for cooling or refrigerating that accomplishes a cooling effect by releasing compressed air through a specially designed opening. Air expands in a rapidly spiraling column of air that separates slow moving molecules (cool) from fast moving molecules (hot).

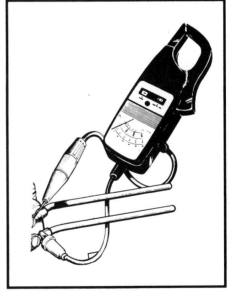

vulcanize: To cure by chemical reaction that induces extensive changes in the physical properties of a rubber or plastic, brought about by reacting it with sulphur and/or other suitable agents; the changes in physical properties include decreased plastic flow, reduced surface tackiness, increased elasticity, much greater tensile strength, and considerably less solubility; the process being hastened by heat and pressure; the method of curing thermosetting materials — rubbers, XLP, etc.

Fig. V-4: One type of testing meter that reads voltage, amperage, and resistance.

VW-1: A UL rating given single conductor cables as to flame resistant properties; formerly FR-1.

W

wall: The thickness of insulation or jacket of cable.

wallboard: A general term applied to any of the many building boards used in place of plaster on interior walls and ceilings.

wall, fire: A dividing wall for the purpose of checking the spread of fire from one part of a building to another.

wall plate: A flush-mounted receptacle or switch plate used to cover the outlet box and wiring device.

wall socket: An electric outlet located in or on the wall for the purpose of providing a source of current.

washer: A small, flat, perforated disk, used to secure the tightness of a joint, screw, etc.

waterblocked cable: A multiconductor cable having interstices filled to prevent water flow or wicking.

water-cooled condenser: Condensing unit that is cooled through the use of water.

waterproof: So constructed that moisture will not interfere with success-ful operation.

water table: A projection of the wall at or near the grade line of a build-ing. It is used to turn the water away from the foundation wall.

watertight: So constructed that water will not enter.

watt: The derived SI unit for power, radiant flux; one watt equals one joule per second.

watt-hour: The number of watts used in one hour.

watt-hour meter: A meter that measures and registers the integral, with respect to time, of the active power in a circuit. A typical watt-hour meter consists of a combination of coils, conductors, and gears—all encased in a housing as shown in Fig. W-1. The coils are constructed on the same principle as a split-phase induction motor, in that the stationary current coil and the voltage coil are placed so that they produce a rotating magnetic field. The disc

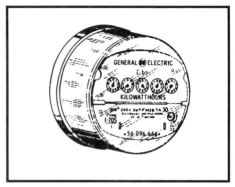

Fig. W-1: A typical watt-hour meter.

near the center of the meter is exposed to the rotating magnetic field. The torque applied to the disc is proportional to the power in the circuit, and the braking action of the eddy currents in the disc makes the speed of rotation proportional to the rate at which the power is consumed. The disc, through a train of gears, moves the pointers on the register dials to record the amount of power used directly in kilowatt hours (kWh).

Most watt-hour meters now have five dials, as shown in Fig. W-2. The dial farthest to the right on the meter counts the kilowatt hours singly. The second dial from the right counts by tens, the third dial by hundreds, the fourth dial from the right by thousands, and the left-hand dial by ten-thousands. Therefore, the reading in the illustration is 2, 2, 1, 7, 9, or 22,179 kWh.

wattmeter: An instrument for measuring the magnitude of the active power in a circuit.

Fig. W-2: The reading of the five dials on this watt-hour meter shows a reading of 2, 2, 1, 7, 9, or 22,179 kWh.

wave: A disturbance that is a function of time or space or both.

waveform: The geometrical shape as obtained by displaying a characteristic of the wave as a function of some variable when plotted over one primitive period.

wavelength: The distance measured along the direction of propagation between two points that are in phase on adjacent waves.

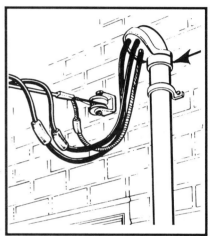

weatherboards: Boards used as an outside covering of buildings, nailed on so as to overlap and shed the rain.

weatherhead: The conduit fitting at a conduit used to allow conductor entry, but prevents weather entry. See Fig. W-3.

weatherproof: So constructed or protected that exposure to the weather will not interfere with successful operation.

web: An open braid; central portion of an I beam.

weight correction factor: The correction factor necessitated by uneven geometric forces placed on cables in conduits; used in computing pulling tensions and sidewall loads.

Fig. W-3: One type of weatherhead used for electric services, and other outdoor applications.

weld: The joining of materials by fusion or recrystallization across the interface of those materials using heat or pressure or both, with or without filler material.

welding cables: Flexible cable used for leads to the rod holders of arc-welders, usually consisting of size 4/0 flexible copper conductors.

welding transformer: A step-down transformer used to produce sufficient instantaneous current to fuse the metals, in contact, through which it is flowing.

Western Union splice: The electrical connection made by paralleling the bared ends of two conductors and then twisting these bared ends, each around the other.

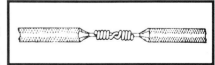

Fig. W-4: Western Union splice.

wet bulb: Device used in measurement of relative humidity.

wet cell battery: A battery having a liquid electrolyte.

wet locations: Exposed to weather or water spray or buried.

Wheatstone bridge: A method of measuring resistances by the proportion existing between the resistance of three known adjustable resistances and the one to be found, all forming the arms of the bridge. See Fig. W-5.

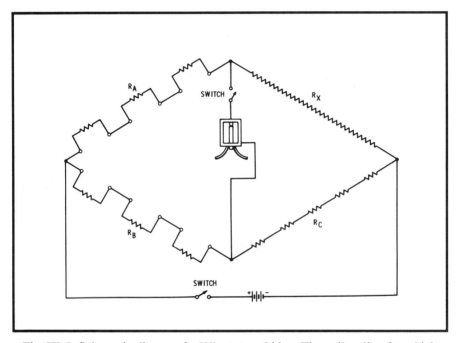

Fig. W-5: Schematic diagrm of a Wheatstone bidge. The coil or line for which the resistance is to be measured is connected as Rx in the bridge circuit. The resistance circuits Ra, Rb, and Rc are known resistances and are called bridge arms.

winch: A machine for hoisting or hauling.

winding: An assembly of coils designed to act in consort to produce a magnetic flux field or to link a flux field.

wing nut: A form of nut that is tightened or loosened by two thin flat wings extending from opposite sides.

wiped joint: A joint wherein filler metal is applied in liquid form and distributed by mechanical action.

wire: A slender rod or filament of drawn metal; the term may also refer to insulated wire.

wire bar: A cast shape that has a square cross section with tapered ends.

wire brush: A hand brush fitted with wire or thin strips of steel instead of bristles. Used for removing rust, dirt, or foreign matter from a surface.

wire, building: That class of wire and cable, usually rated at 600V, which is normally used for interior wire of buildings.

wire, covered: A wire having a covering not in conformance with NEC standards for material or thickness.

wire drawing: The process by which wire is made; as by drawing metal through a hole in a steel plate.

wire, hookup: Insulated wire for low voltage and current in electronic circuits.

wire, resistance: Wire having appreciable resistance; used in heating applications such as electric toasters, heaters, etc.

WM: Wattmeter.

work: Force times distance: pound-feet.

work function: The minimum energy required to remove an electron from the Fermi level of a material into field-free space; units — electron volts.

work hardening: Hardening and embrittlement of metal due to cold working.

working drawing: A drawing containing all dimensions and instructions necessary for successfully carrying a job to completion. See Fig. T-6 on page 216.

wrench: A tool for exerting a twisting strain, as in tightening a nut or bolt.

Wye connection: Three-phase transformer connection wherein all three phases are connected at a central point, forming a "Y" configuration. See Y-Delta connection and Y-Y connection on pages 219 and 220.

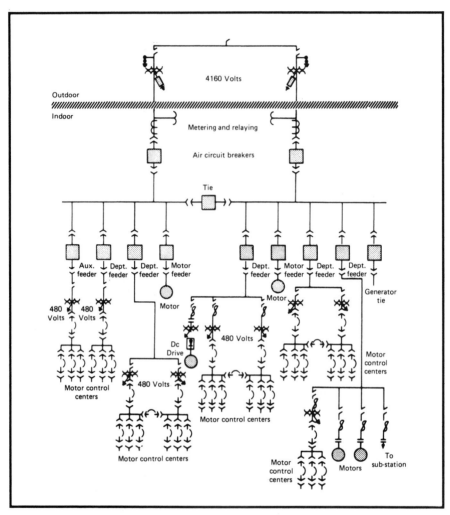

Fig. W-6: Sample drawing sheet from a set of electrical working drawings.

X

xenon: A chemical element, Xe, atomic number 54. It is a member of the family of noble gases, group O in the periodic table. Its main use is to fill a type of flashbulb employed in photography and called electronic speed light.

Xmfr: Abbreviation for transformer. Used mainly on electrical drawings. See Fig. X-1.

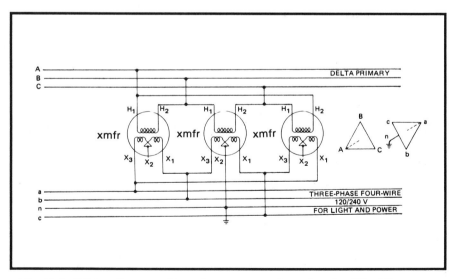

Fig. X-1: Wiring diagram of transformer connections showing "Xmfr" to designate the transformers to save drafting time.

x-ray: 1) Penetrating short wavelength electromagnetic radiation created by electron bombardment in high voltage apparatus; produce ionization when the rays strike certain materials. The rays pass through most objects as though they were transparent. 2) A popular name for Roentgen rays. A form of radiant energy sent out when the cathode rays of a Crookes tube strike upon the opposite walls of a tube or upon any object in the tube.

x-unit: A unit of length, approximately 10^{-11} cm, formerly used for measuring x-ray wavelengths and crystal dimensions.

xylene: One of a group of three isometric aromatic hydrocarbons. The three isometric sylenes and ethylbenzene all have in common a molecular weight of 106.2 and the simplified formula, C_8H_{10}.

Orginally designated "coal tar" hydrocarbons, these compounds have, since World War II, been almost exclusively produced from petroleum by hydroforming or catalytic reforming of appropriate naphtha fractions.

Y

yardage: Relates to cubic yards of earth excavated.

Y-Delta Connection: A three-phase transformer connection wherein the primary phases are connected at a central point and the secondary phases are connected in series with each other. See Fig. Y-1.

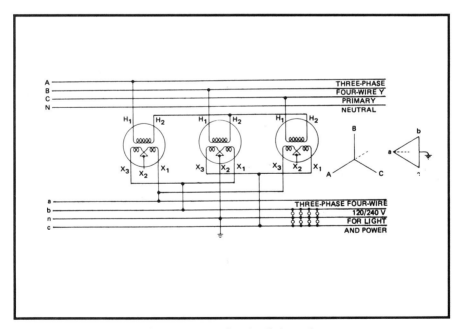

Fig. Y-1: Y-Delta transformer connection for light and power.

yield point: That unit stress at which the specimen begins to stretch without increase in the load.

yield strength: 1) The point at which a substance changes from elastic to viscous. 2)The load required to produce a permanent stretch or elongation of a material.

yield value: The lowest pressure at which a plastic will flow. Below this pressure the plastic behaves as an elastic solid; above this pressure as a viscous liquid.

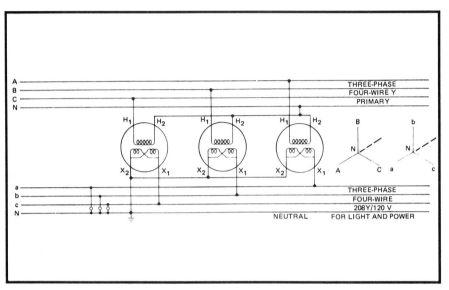

Fig. Y-2: Y-Y transformer connection for light and power.

yoke: 1) The ferromagnetic ring or cylinder that holds the pole pieces of a dynamo-type generator and acts as a part of the magnetic circuit. 2) The system of coils used for magnetic deflection of the electron beam in cathode-ray tubes, such as television picture tubes.

Y-Y Connection: A three-phase transformer connection whereas the three primary phases are connected at a central point. The three secondary phases are also connected at a central point. See Fig. Y-2.

Z

zero: The numeral 0; a cipher; the lowest point. For example, "the voltmeter shows a reading of zero," meaning no voltage is present.

zig-zag connection: A connection of polyphase windings in which each branch generates phase-displaced voltages.

zinc: A bluish-white metallic element with a symbol of Zn, the atomic number 30, and the atomic weight 65.38. Zinc is familiar as the negative electrode material in dry cells and as a protective coating for some sheet metals used in electronics.

zinc-alkaline battery: A relatively new type of storage battery introduced some years ago. Due to its high cost, it has gained very little commercial appeal. However, space projects are utilizing this type of battery due to its greater power output compared to its size and weight. Compared with the conventional lead-acid cell, the zinc cell has a watt-hour output per pound of cell over five times the output of the lead cell, while its watt-hour output per cubic inch of cell is three to four times as much.

zinc-carbon battery: A primary cell in which the negative electrode is zinc and the positive electrode is carbon, and which may be either wet or dry. See Fig. Z-1.

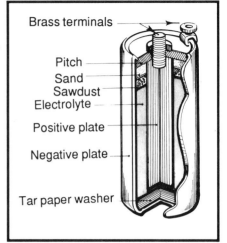

zinc chloride (ZnCl₂): A white deliquescent salt, obtained by the solution of zinc or zinc oxide, in hydrochloric acid, or by burning zinc in chlorine. Used as a soldering flux.

zonal cavity method: A calculation method used to determine the amount of illumination produced by a given installation of lighting fixtures in a particular room or other indoor area.

Fig. Z-1: Principal parts of a zinc-carbon battery.

The zonal cavity method of calculating average illumination levels assumes each room or area to consist of the following three separate cavities: ceiling cavity, room cavity, and floor cavity.

Figure Z-2 on page 223 shows that the *ceiling cavity* extends from the lighting fixture plane upward to the ceiling. The *floor cavity* extends from the work plane downward to the floor, while the *room cavity* is the space between the lighting fixture plane and the working plane.

If the lighting fixtures are recessed or surface-mounted on the ceiling, there will be no ceiling cavity and the ceiling cavity reflectance will be equal to the actual ceiling reflectance. Similarly, if the work plane is at floor level, there will be no floor cavity and the floor cavity reflectance will be equal to the actual floor reflectance. The geometric proportions of these spaces become the "cavity ratios."

Cavity ratio: Rooms are classified according to shape by *ten* cavity ratio numbers. The basic formula for obtaining cavity ratios in rectangular-shaped rooms is:

$$Cavity\ Ratio = \frac{5 \times Height\,(Length + Width)}{Length \times Width}$$

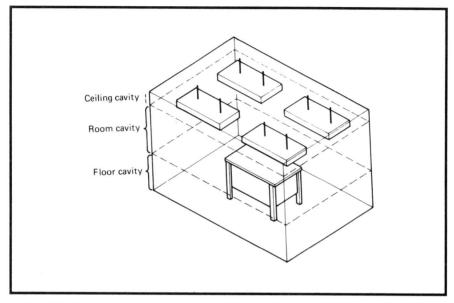

Fig. Z-2: The three room cavities as applied to the zonal cavity method of lighting calculations.

where height is the height of the cavity under consideration—that is, ceiling, floor, or room cavity.

The zonal cavity method of making lighting calculations provides greater flexibility with increased accuracy over any previous method of determining average illumination levels.

There are five key steps in using the zonal-cavity method:

- Determine the required level of illumination.

- Determine the coefficient of utilization.

- Determine the maintenance factor.

- Calculate the number of lamps and lighting fixtures required.

- Determine the location of the lighting fixtures.

The general applications of the zonal cavity method are to determine the number of lighting fixtures that are required to produce a given lighting level in footcandles and to determine what lighting level will be produced by a given number of lighting fixtures.